Vectors to Velocity: A Guide to Motion and Mechanics   By: Prashant Kumar Lal

Vectors to Velocity

A Guide to Motion and Mechanics

BY

PRASHANT KUMAR LAL

Vectors to Velocity: A Guide to Motion and Mechanics

BY

PRASHANT KUMAR LAL

Copyright © Prashant Kumar Lal, 2024 - Author

No part of this publication may be reproduced, distributed, or transmitted in any form or by any means, including photocopying, recording, or other electronic or mechanical methods, without the prior written permission of the author, except in the case of brief quotations embodied in critical reviews and certain other non-commercial uses permitted by copyright law. For permission requests, write to the author at the address provided in this publication.

First Edition, 2024

ISBN: 9798346652540

Imprint: Independently published

Imprint: Independently published

Published By: **Prashant Kumar Lal**
R – 133, Phase 2, Sarabha Nagar Extension
Pakhowal Road, Ludhiana (Pb), Pin – 142022, (INDIA)
Email: prasshantlal1961@gmail.com,
Phone: +91 9478624982

Publishing Fields: Science & Technology, Educational Books

About the Author

Prashant Kumar Lal is a distinguished educator, author, and retired Principal with over 38 years of experience in education, including 20 years as a school head. Known for his commitment to teaching and his passion for Physics, he has dedicated decades to guiding students, teachers, and schools towards excellence. His career in education is marked by a deep-seated commitment to helping students grasp complex subjects and develop a genuine love for learning.

He became a pioneer in school administration and teaching methods in India, drawing on his extensive experience to provide consultancy services through his company,

Vectors to Velocity: A Guide to Motion and Mechanics   By: Prashant Kumar Lal

Prashant Educational Consultancy Services OPC Pvt Ltd. His consultancy work supports schools in administrative and academic processes, teacher training, and career guidance, helping them align with contemporary educational standards.

Prashant Kumar Lal is also a prolific writer, with several books to his credit, including "Speeches from the Desk of the Principal," "Physics Test Series for Class XII," "The Legend of Inara Wali," and his highly regarded poetry collection, "Image of My Experiences." These works reflect his varied interests in literature, science, and philosophy, as well as his talent for conveying complex ideas with clarity and insight.

With his latest book, **"Vectors to Velocity: A Guide to Motion and Mechanics,"** Prashant Kumar Lal continues his lifelong mission to make Physics accessible and engaging for students. His extensive experience and passion for education shine through each page, making this book an invaluable resource for students aspiring to excel in Physics and develop a lasting interest in the subject.

∽∞✕✕∞∾

About the Book

"Vectors to Velocity: A Guide to Motion and Mechanics" by Prashant Kumar Lal is an insightful guide crafted to support students in mastering the core concepts of kinematics and Newton's laws of motion, aligning with the CBSE Class XI Physics curriculum. This book is designed to simplify complex topics, making them accessible for students who may find Physics challenging, while also engaging those who wish to delve deeper into the principles of motion and mechanics.

This comprehensive guide explores the journey from foundational vector concepts to the intricate dynamics of motion, acceleration, and force, with a clear and structured approach. Prashant Kumar Lal, a seasoned educator in Physics, combines his expertise with a passion for teaching to present these concepts in a straightforward, student-friendly manner.

Key Features:

1. Clear and Concise Explanations: Each topic is broken down with clarity, helping students understand concepts such as vectors, velocity, acceleration, and Newton's laws. Lal uses practical examples, relatable analogies, and step-by-step guidance to demystify these fundamental areas.

2. Illustrative Diagrams and Visuals: To aid comprehension, the book includes numerous diagrams and illustrations, providing visual representations that simplify complex ideas. Motion graphs, vector diagrams, and real-world examples of forces and trajectories make the subject matter more accessible.

3. Real-World Applications: Lal connects Physics principles to everyday experiences, demonstrating the relevance of kinematics and mechanics. From sports to space travel, these practical insights encourage students to see the physics behind real-world phenomena.

4. Problem-Solving Techniques: A dedicated section on problem-solving strategies helps students build analytical and critical thinking skills, covering a variety of questions from basic exercises to more advanced applications. This focus equips students to handle both exam questions and practical problem-solving.

5. Historical and Conceptual Context: Introducing some of the historical developments and scientific figures behind these discoveries, the book offers a broader perspective,

enhancing students' understanding of the evolution of scientific thought.

6. Alignment with CBSE Standards: Every chapter is crafted to meet and exceed CBSE requirements, ensuring thorough preparation for examinations. The book's structure and content are specifically designed to align with the curriculum and help students develop a solid foundation.

"Vectors to Velocity" serves as a valuable learning tool and reference for Class XI Physics students, whether they are preparing for exams or pursuing a deeper understanding of the subject. Prashant Kumar Lal's extensive experience and dedication to teaching Physics resonate throughout, making this book both an academically enriching and engaging resource for students.

XXXX

Vectors to Velocity: A Guide to Motion and Mechanics   *By: Prashant Kumar Lal*

BOOKS BY THE SAME AUTHOR

1. Science Unfolded for Std VI
2. Science Unfolded for Std VII
3. Science Unfolded for Std VIII
4. Physics Test Series for class XII
5. Image of my Experiences: A book of English Poetry
6. Speeches from the Desk of the Principal
7. The Legend of Inara Wali
8. My Pen and My Universe (Chronicles of Life, Love and Learning)
9. Fables and Fantasies: A World of Stories for Children
10. The Physical World: Building Blocks of Physics: For Class XI
11. Velocity to Vectors: A Guide to Motion and Mechanics

CXXXYO

Preface

Physics is the study of the world around us, from the smallest particles to the vastness of space. It explores the forces, motions, and principles that govern everything in our universe. With "Vectors to Velocity: A Guide to Motion and Mechanics," I aim to make these concepts both accessible and engaging for students beginning their journey in Physics.

This book is designed to align with the Class XI CBSE curriculum, focusing on foundational topics like vectors, kinematics, and Newton's laws of motion. My goal is to break down these topics in a way that not only prepares students for exams but also inspires them to look at the world with curiosity and an analytical mind. Through clear explanations, practical examples, and illustrative diagrams, I have strived to make complex ideas easy to understand, even for those who may find Physics challenging at first.

My experience as a teacher and school principal has shown me the importance of connecting with students and addressing their unique learning needs. This book reflects years of teaching, during which I have learned the value of clarity, patience, and a structured approach to education. Alongside academic content, I have included problem-solving techniques, real-world applications, and insights into the history of Physics to help students see both the practicality and beauty of the subject.

I hope that "Vectors to Velocity" will serve as a reliable resource for students, supporting their academic journey and fostering a lifelong appreciation for Physics. It is my wish that this book will empower students to tackle challenges, ask questions, and discover the joy of learning about the physical world.

Thank you for choosing to explore this book. I hope it becomes a valuable part of your educational journey.

— Prashant Kumar Lal, Author

❁XXXX❁

Vectors to Velocity: A Guide to Motion and Mechanics *By: Prashant Kumar Lal*

INDEX

1. Kinematics: Motion in a Straight Line ---
Page 16- 128

- Frame of Reference
- Concept of point and extended objects
- Types of frames (inertial and non-inertial)
- Position, path length, and displacement
- Average velocity and average speed
- Instantaneous velocity and speed
- Acceleration
- Graphical representation of motion
- Uniform and non-uniform motion, uniform acceleration

- Equations of motion (graphical treatment)

2. Motion in a Plane
---Page 129 - 255

- Scalar and vector quantities, position and displacement vectors

- Equality of vectors, multiplication of vectors by a real number; addition and subtraction of vectors

- Relative velocity

- Motion in a plane, cases of uniform velocity and uniform acceleration, projectile motion

- Uniform circular motion

◦✗✗✗✗◦

Chapter – 1

KINEMATICS

1.1 Introduction

What is Kinematics?

Understanding Motion: The Study of Kinematics

Welcome to the fascinating world of physics! As we embark on this journey, we'll begin by exploring the fundamental concepts of motion. Kinematics, the study of motion without considering forces, is the cornerstone of understanding how objects move and interact with their surroundings.

In our daily lives, we observe various types of motion – from the simplest actions like walking or throwing a ball to complex phenomena like planetary orbits and rocket trajectories. Kinematics provides a systematic approach to describing and analysing these motions, enabling us to predict and understand the world around us.

Kinematics is the branch of physics that deals with the study of motion, focusing on the description of an object's position, displacement, velocity, acceleration, and time. It provides a mathematical framework for understanding how objects move, without considering the underlying forces that cause this motion. with the study of motion, focusing on the description of an object's position, displacement, velocity, acceleration, and time. It provides a mathematical framework for understanding how objects move, without considering the underlying forces that cause this motion.

1. Analyse complex motions in various contexts.

2. Predict the behaviour of objects under different conditions.

3. Design and optimize systems involving motion.

Learning Objectives

After completing this chapter, you'll be able to:

1. Describe motion using kinematic terminology.

2. Apply kinematic equations to solve problems.

3. Interpret graphical representations of motion.

4. Analyse relative motion and reference frames.

5. Solve problems involving various types of motion.

Let's Get Started!

In the following sections, we'll explore the concepts of kinematics in detail, using examples, illustrations, and practice problems to reinforce your understanding. Buckle up and join me on this exciting journey into the world of motion!

Kinematics is the study of motion without considering forces. It describes the motion of objects, including their position, displacement, velocity, acceleration, and time.

1.2 Description of Motion

To describe motion, we use the following terms:

- **Position**: *The location of an object in space. Position: The Foundation of Kinematics*

In kinematics, position is the fundamental concept that describes where an object is located in space. It's the starting point for understanding motion, and it's essential to grasp this concept before diving deeper into kinematics.

What is Position?

Position refers to the location of an object in relation to a reference point or frame of reference. It's a vector quantity, meaning it has both magnitude (amount of movement) and direction.

Key Concepts:

1. **Reference Point**: A fixed point from which position is measured.

O P

Vectors to Velocity: A Guide to Motion and Mechanics   By: Prashant Kumar Lal

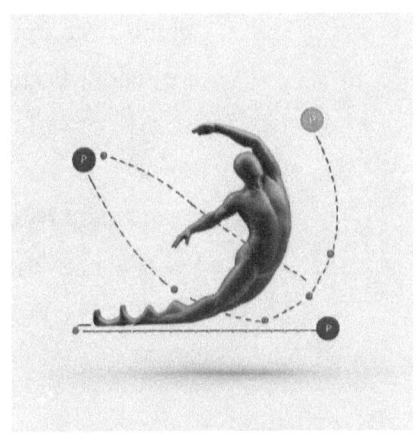

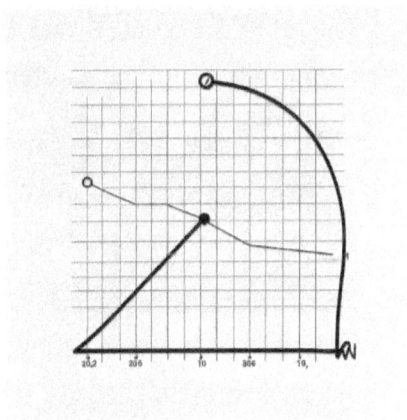

Key Components:

a) Reference Point (O): The fixed point from

which measurements are made. Point 'O' is called 'Reference Point'.
  b) Coordinate System (x, y): A grid system used to locate points in space

Description:

- The reference point (O) serves as the origin of the coordinate system.

- The object (P) is located at coordinates (x, y) relative to the reference point.

- Measurements are made from the reference point to determine the object's position.

Example:

Suppose we want to locate a book on a shelf. We can use the top-left corner of the shelf as the reference point (O).

The book's position can be measured as (x, y) coordinates from this point.

2. **Frame of Reference**: A coordinate system that defines the orientation of the reference point.

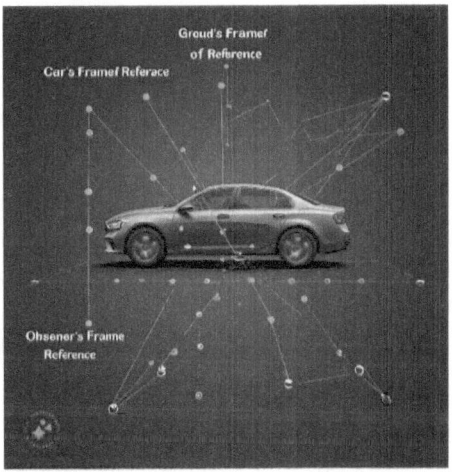

In kinematics, understanding motion requires a clear perspective. That's where the frame of reference comes in – a crucial concept that helps us describe and analyse motion.

What is a Frame of Reference?

A frame of reference is a coordinate system that defines the orientation and position of an object or event. It's a mental framework that allows us to:

1. Describe motion

2. Measure position, velocity, and acceleration

3. Analyse and interpret data

Types of Frames of Reference:

1. Inertial Frame: A stationary or non-accelerating frame (e.g., Earth's surface).

2. Non-Inertial Frame: An accelerating frame (e.g., a moving car).

Key Components:

1. Origin (O): The reference point (0, 0) of the coordinate system.

2. Axes: x, y, and z axes define the directions.

3. Coordinates: Numbers that locate points in space.

Examples:

1. GPS Navigation: Uses Earth as a reference frame.

2. Sports: The playing field serves as a reference frame.

3. Space Exploration: Stars and planets provide a reference frame

Importance:

1. Relative Motion: Frames of reference help understand relative motion.

2. Motion Description: Accurate description of motion requires a clear frame.

3. Problem-Solving: Frames simplify complex motion problems.

Real-World Applications:

1. Navigation Systems: GPS, aircraft navigation.

2. Robotics: Precise motion control.

3. Physics Experiments: Accurate measurements.

Common Frames of Reference:

1. Cartesian Coordinates: x, y, z axes.

2. Polar Coordinates: r, θ (radius, angle).

3. Spherical Coordinates: r, θ, φ (radius, polar angle, azimuthal angle).

Tips and Tricks:

1. Choose a convenient origin.

2. Align axes with motion.

3. Consistency is key.

3. **Coordinate System**: A system used to locate points in space (e.g., Cartesian, polar).

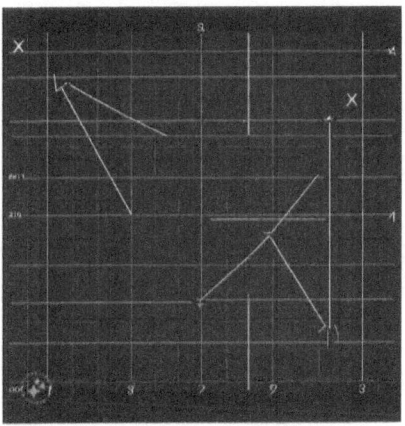

Types of Position:

1. Initial Position: The starting position of an object.

2. Final Position: The ending position of an object.

3. Displacement: The change in position (more on this later).

Measuring Position:

Position can be measured using various units, including:

1. Meter (m): Standard unit for length.

2. Centimetre (cm): 1/100 of a meter.

3. Kilometre (km): 1000 meters.

Graphical Representation:

Position can be represented graphically using:

1. Number Lines: A one-dimensional representation.

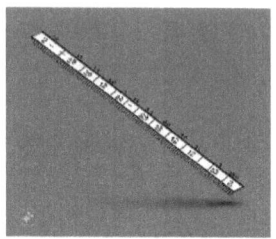

2. Coordinate Grids: A two-dimensional representation.

Vectors to Velocity: A Guide to Motion and Mechanics　　　　By: Prashant Kumar Lal

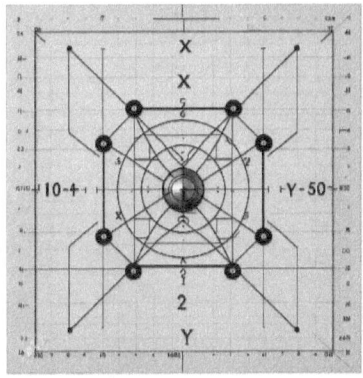

3. Vector Diagrams: Show direction and magnitude.

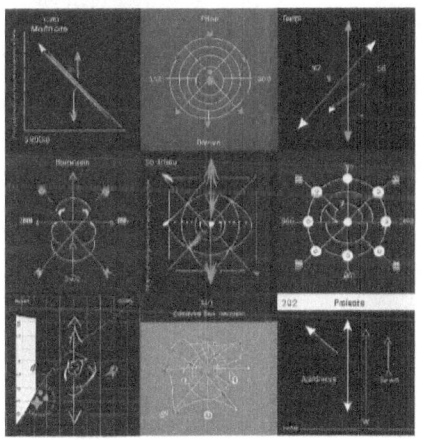

Real-World Examples:

1. GPS Navigation: Uses position to guide vehicles.

2. Sports: Position tracking in games like football, basketball.

3. Robotics: Position sensors for precise movement.

Equations and Formulas:

1. **Position-Time Equation**:
   $x(t) = x_0 + v_0 t + (1/2) at^2$
   he equation you're referring to is the equation of motion for an object under constant acceleration, which is a fundamental concept in physics. Here's how to derive it:

   Derivation:
   Let's consider an object moving along a straight line (x-axis) with initial position x_0, initial velocity v_0, and constant acceleration a.

Step 1: Define the acceleration

$a = dv/dt$ (acceleration is the rate of change of velocity)

Step 2: Integrate acceleration to find velocity

$v(t) = \int a \, dt = v_0 + at$ (integrate acceleration to find velocity)

Step 3: Integrate velocity to find position

$x(t) = \int v(t) \, dt = \int (v_0 + at) \, dt$

Step 4: Evaluate the integral

$x(t) = x_0 + v_0 t + (1/2) at^2$

Equation of Motion:

$x(t) = x_0 + v_0 t + (1/2) at^2$
where:

$x(t)$ = position at time t

x_0 = initial position

v_0 = initial velocity

a = constant acceleration

t = time

This equation describes the position of an object under constant acceleration, which is essential in understanding various phenomena in physics, engineering, and other fields.

2. **Displacement Equation**:

$$\Delta X = X_f - X_i$$

This equation calculates the change in position, or displacement, of an object.

Variables:

- ΔX (Delta X): Displacement (change in position)

- X_f (X final): Final position

- X_i (X initial): Initial position

Explanation:

Displacement is a vector quantity that measures the distance between an object's initial and final positions. It's the shortest distance between the two points, and its direction is from the initial to the final position.

Example:

A ball moves from point A ($X_i = 5$ m) to point B ($X_f = 12$ m).

$\Delta X = X_f - X_i$

$= 12$ m $- 5$ m

$= 7$ m

The ball's displacement is 7 meters.

Key Points:

1. Displacement is different from distance travelled. Distance travelled is the total length of the path, while displacement is the straight-line

distance between the initial and final positions.

2. Displacement can be positive or negative, depending on the direction of motion.

3. If the object returns to its initial position, the displacement is zero.

Graphical Representation:

Imagine a number line:

...----- 0 -----======>

5 (X_i) ======> 12 (X_f)

$\Delta X = X_f - X_i = 12 - 5 = 7$

The displacement equation calculates the difference between the final and initial positions, giving you the change in position.

Practice Problems:

1. Find the position of an object after 5 seconds, given an initial velocity of 2 m/s and acceleration of 3 m/s^2.

2. Calculate the displacement of an object moving from (3, 4) to (6, 8) on a coordinate grid.

Summary:

Position is the foundation of kinematics, describing an object's location in space. Understanding position is crucial for analysing motion, displacement, velocity, and acceleration.

Key Terms:

- Position

- Reference point

- Frame of reference

- Coordinate system

- Initial position

- Final position

- Displacement

Points to Remember:

- Position is a vector quantity.

- Reference points and frames of reference are essential.

- Coordinate systems simplify position measurement

- Displacement: The change in position of an object.

- Distance: The total path covered by an object.

- Velocity: The rate of change of displacement.

- Acceleration: The rate of change of velocity.

Average Speed and Speed

Introduction

Speed and average speed are fundamental concepts in physics that help us understand motion. In this chapter, we'll explore the differences between speed and average speed, and learn how to calculate them.

Speed

Speed is the rate of change of an object's position with respect to time. It's a scalar quantity, measured in meters per second (m/s) or kilometres per hour (km/h).

Formula:

Speed (v) = Distance (s) / Time (t)

$v = s / t$

Units:

m/s, km/h, mph (miles per hour)

Example 1:

A car travels 100 m in 10 s. Find its speed.

$v = s / t$

$= 100 / 10$

$= 10$ m/s

Average Speed

Average speed is the total distance travelled divided by the total time taken. It's also a scalar quantity.

Formula:

Average Speed (V_{avg}) = Total Distance (s) / Total Time (t)

$V_{avg} = s / t$

Units:

m/s, km/h, mph

Example 2:

A cyclist travels 50 km in 2 hours and then 30 km in 1 hour. Find the average speed.

Total Distance = 50 + 30 = 80 km

Total Time = 2 + 1 = 3 hours

$V_{avg} = s / t$

= 80 / 3

= 26.67 km/h

Key Differences:

1. Speed is instantaneous, while average speed is over a period.

2. Speed can vary, but average speed remains constant.

Problems and Exercises

Multiple Choice Questions

1. What is the unit of speed?

A) m/s

B) km/h

C) Both A and B

D) None

Answer: C) Both A and B

2. Which is the formula for average speed?

A) $V_{avg} = s / t$

B) $V_{avg} = s + t$

C) $V_{avg} = s - t$

D) $V_{avg} = s \times t$

Answer: A) $V_{avg} = s / t$

Short Answer Questions

1. Define speed and average speed.

Answer: Speed is the rate of change of an object's position with respect to time, while average speed is the total distance travelled divided by the total time taken.

2. Calculate the speed of a car traveling 200 m in 20 s.

Answer: $v = s / t = 200 / 20 = 10$ m/s

Numerical Problems

1. A train travels 500 km in 5 hours. Find its average speed.

Answer: $V_{avg} = s / t = 500 / 5 = 100$ km/h

2. A cyclist travels 20 km in 2 hours and then 10 km in 1 hour. Find the average speed.

Answer: Total Distance = 20 + 10 = 30 km

Total Time = 2 + 1 = 3 hours

V_{avg} = s / t = 30 / 3 = 10 km/h

1. A car travels 100 m in 5 s. Find its speed.

Answer: v = s / t = 100 / 5 = 20 m/s

Long Answer Questions

1. Explain the difference between speed and average speed.

Answer: Speed is instantaneous, while average speed is over a period. Speed can vary, but average speed remains constant.

2. Calculate the average speed of a journey from city A to city B, covering 200 km in 4 hours, and then 150 km in 2 hours.

Answer: Total Distance = 200 + 150 = 350 km

Total Time = 4 + 2 = 6 hours

V_{avg} = s / t = 350 / 6 = 58.33 km/h

Hints and Tips:

1. Always check units.

2. Use formulas correctly.

3. Break down complex problems into simpler ones.

By mastering speed and average speed, you'll better understand motion and be able to solve complex problems in physics. Practice regularly and remember to check your units!

Instantaneous Velocity and Speed

Introduction

In physics, understanding motion is crucial, and velocity plays a vital role. Instantaneous velocity and speed are fundamental concepts that help us analyse an object's motion at a specific point. In this chapter, we'll explore the differences between instantaneous velocity and speed, and learn how to calculate them.

Instantaneous Velocity

Instantaneous velocity is the velocity of an object at a specific instant. It's a vector quantity, measured in meters per second (m/s) or kilometres per hour (km/h).

Formula:

Instantaneous Velocity

$(v) = \text{Limit} \ (\Delta s / \Delta t)$ as $\Delta t \to 0$

$v = ds/dt$

Units:

m/s, km/h, mph (miles per hour)

Example 1: A car travels 100 m in 10 s. Find its instantaneous velocity at t = 5 s.

v = ds/dt

= (100 m) / (10 s)

= 10 m/s (at t = 5 s)

Instantaneous Speed

Instantaneous speed is the magnitude of instantaneous velocity. It's a scalar quantity.

Formula:

Instantaneous Speed

$(|v|) = \sqrt{(v_x^2 + v_y^2)}$

Units:

m/s, km/h, mph

Example 2:

A particle moves with velocity 3i + 4j m/s. Find its instantaneous speed.

$|v| = \sqrt{(3^2 + 4^2)}$

$= \sqrt{(9 + 16)}$

$= \sqrt{25}$

$= 5$ m/s

Key Differences:

1. Instantaneous velocity is a vector, while instantaneous speed is a scalar.

2. Instantaneous velocity has direction, while instantaneous speed does not.

Numerical Problems

1. A car travels 200 m in 20 s. Find its instantaneous velocity at t = 10 s.

Answer: v = ds/dt = (200 m) / (20 s) = 10 m/s

2. A particle moves with velocity 2i - 3j m/s. Find its instantaneous speed.

Answer: $|v| = \sqrt{(2^2 + (-3)^2)} = \sqrt{(4 + 9)} = \sqrt{13} \approx 3.61$ m/s

Questions with Hints

Multiple Choice Questions

1. What is the unit of instantaneous velocity?

A) m/s

B) km/h

C) Both A and B

D) None

Hint: Think about velocity units.

Answer: C) Both A and B

2. Which is the formula for instantaneous speed?

A) $|v| = \sqrt{(v_x^2 + v_y^2)}$

B) $|v| = v_x + v_y$

C) $|v| = v_x - v_y$

D) $|v| = v_x \times v_y$

Hint: Recall Pythagorean theorem.

Answer: A) $|v| = |v| = \sqrt{(v_x^2 + v_y^2)}$

Short Answer Questions

1. Define instantaneous velocity and speed.

Hint: *Explain vector and scalar quantities.*

Answer: Instantaneous velocity is the velocity at a specific instant, while instantaneous speed is its magnitude.

2. Calculate the instantaneous velocity of a car traveling 300 m in 30 s.

Hint: Use $v = ds/dt$.

Answer: $v = (300 \text{ m}) / (30 \text{ s}) = 10 \text{ m/s}$

Long Answer Questions

1. Explain the difference between instantaneous velocity and speed.

Hint: Discuss vector and scalar quantities.

Answer: Instantaneous velocity has direction, while instantaneous speed does not.

2. Calculate the instantaneous speed of a particle moving with velocity 4i + 5j m/s.

Hint: Use $|v| = \sqrt{(v_x^2 + v_y^2)}$.

Answer: $|v| = \sqrt{(4^2 + 5^2)} = \sqrt{(16 + 25)} = \sqrt{41} \approx 6.4$ m/s

Hints and Tips:

1. Always check units.

2. Use formulas correctly.

3. Break down complex problems into simpler ones.

4. Visualize motion graphs.

Mastering instantaneous velocity and speed will help you analyse motion accurately. Practice regularly and remember to check your units!

1.3 Types of Motion

There are several types of motion:

- Rectilinear Motion: Motion along a straight line.

- Circular Motion: Motion along a circular path.

- Rotational Motion: Motion around a fixed axis.

- Oscillatory Motion: Motion that repeats itself.

1.4 Kinematic Equations

The following equations relate position, velocity, acceleration, and time:

- First Equation: $v = u + at$

The equation:

$$v = u + at$$

is the fundamental equation of motion under constant acceleration, where:

- v = final velocity

- u = initial velocity

- a = constant acceleration

- t = time

Derivation:

Consider an object moving with initial velocity u and accelerating uniformly at rate a.

Step 1: Define acceleration

$a = \Delta v / \Delta t$ (acceleration is the rate of change of velocity)

Step 2: Rearrange the equation

$\Delta v = a \Delta t$

Step 3: Substitute $\Delta v = v - u$ (change in velocity)

$v - u = a\Delta t$

Step 4: Solve for v

$v = u + at$

Therefore, the equation $v = u + at$ represents the relationship between initial and final velocities under constant acceleration.

Physical Interpretation:

1. If $a > 0$, the object accelerates (speeds up).

2. If $a < 0$, the object decelerates (slows down).

3. If $a = 0$, the object moves with constant velocity ($u = v$).

Example:

A car accelerates from 20 m/s to 40 m/s in 4 seconds. Find its acceleration.

u = 20 m/s, v = 40 m/s, t = 4 s

a = (v - u) / t

= (40 - 20) / 4

= 20 / 4

= 5 m/s²

The car accelerates at 5 m/s

- **Second Equation:**

$s = ut + (1/2)\, at^2$

The equation:

s = ut + (1/2) at²

is the equation of motion under constant acceleration, where:

- s = displacement

- u = initial velocity

- t = time

- a = constant acceleration

Derivation:

Step 1: Define displacement

Displacement (s) is the integral of velocity (v) with respect to time (t).

$s = \int v \, dt$

Step 2: Substitute v = u + at (from previous derivation)

$s = \int (u + at) \, dt$

Step 3: Integrate

$s = ut + (1/2) \, at^2 + C$

where C is the constant of integration.

Step 4: Apply initial condition

At t = 0, s = 0 (object starts at initial position).

$0 = u(0) + (1/2) a (0)^2 + C$

$C = 0$

Step 5: Simplify

$s = ut + (1/2) at^2$

Physical Interpretation:

1. The first term (u t) represents the displacement due to initial velocity.

2. The second term $((1/2) at^2)$ represents the displacement due to acceleration.

Example:

A car starts from rest (u = 0) and accelerates at 2 m/s² for 4 seconds.

$s = 0(4) + (1/2)(2)(4)^2$

$= 0 + (1/2)(2)(16)$

$= 16 \text{ m}$

The car displaces 16 meters in 4 seconds.

- Third Equation: $v^2 = u^2 + 2as$

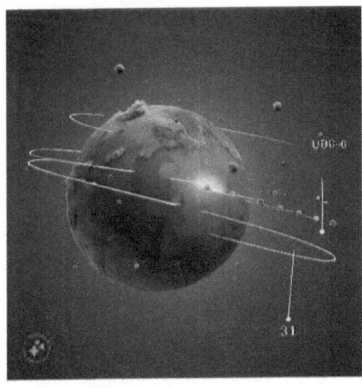

The equation:

$$v^2 = u^2 + 2as$$

is a fundamental equation of motion under constant acceleration.

Derivation:

Step 1: Start with the equation $v = u + at$

Step 2: Square both sides

$$v^2 = (u + at)^2$$

Step 3: Expand the right side

$v^2 = u^2 + 2uat + a^2t^2$

Step 4: Recall the equation $s = ut + (1/2)at^2$

Step 5: Multiply both sides by 2a

$2as = 2uat + a^2t^2$

Step 6: Substitute this expression into the expanded equation

$v^2 = u^2 + 2as$

Therefore, $v^2 = u^2 + 2as$.

Physical Interpretation:

1. Kinetic energy: v^2 is proportional to kinetic energy.

2. Work-energy theorem: $2as$ represents the work done by the force ($F = ma$).

Example:

A car accelerates from 10 m/s to 20 m/s over 50 m.

$u = 10$ m/s, $v = 20$ m/s, $s = 50$ m

$2as = v^2 - u^2$

$2a(50) = 20^2 - 10^2$

$100a = 300$

$a = 3$ m/s²

The car accelerates at 3 m/s².

1.5 Graphical Analysis

Graphs help visualize motion:

- **Position-Time Graph: Shows position vs. time**.

A position-time graph is a graphical representation of an object's position as a function of time. It's a fundamental tool in physics to visualize and analyse motion.

Key Features:

1. Time (t) is plotted on the x-axis.

2. Position (s) or displacement (x) is plotted on the y-axis.

3. The graph shows the object's position at various times.

Types of Motion on a Position-Time Graph:

1. Straight line parallel to x-axis: Object at rest (zero velocity).

2. Straight line with slope: Uniform motion (constant velocity).

3. Curved line: Non-uniform motion (acceleration).

4. zig-zag or oscillating line: Periodic motion (vibration or oscillation).

Importance:

1. Visualizes motion: Helps understand complex motions.

2. Analyses velocity and acceleration: Slope and curvature indicate velocity and acceleration.

3. Identifies patterns: Periodic, oscillatory, or chaotic behaviour.

4. Compares motions: Different objects or scenarios.

5. Solves problems: Determines position, velocity, and acceleration at specific times.

6. Real-world applications: Trajectory planning (e.g., GPS), motion control systems, and physics simulations.

Interpretation:

1. Slope ($\Delta s/\Delta t$) represents velocity.

2. Curvature represents acceleration.

3. Intersection with x-axis indicates return to initial position.

4. Asymptotes indicate constant velocity or position.

Real-World Examples:

1. Tracking satellite orbits.

2. Modelling population growth or decline.

3. Analysing sports performance (e.g., sprinter's velocity).

4. Understanding mechanical systems (e.g., piston motion).

By analysing position-time graphs, you can gain insights into various physical phenomena, making it an essential tool for physicists, engineers, and scientists.

Time (s)	0	2	4	6	8
(Position (m)	0	5	10	15	20

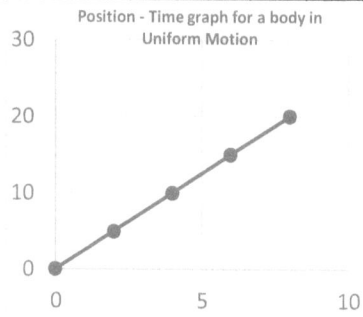

Position - Time graph for a body in Uniform Motion

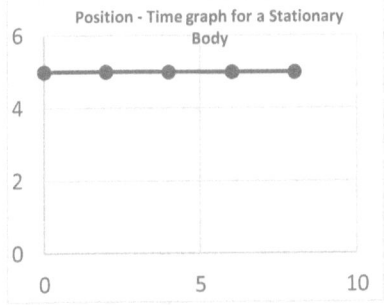

Time (s)	0	2	4	6	8
(Position (m)	5	5	5	5	5

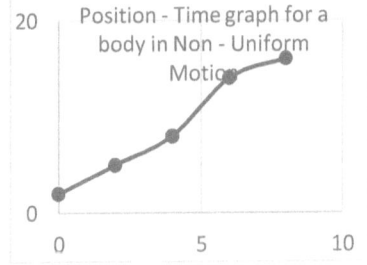

Time (s)	0	2	4	6	8
(Position (m)	2	5	8	14	16

- **Velocity-Time Graph: Shows velocity vs. time.**

A velocity-time graph is a graphical representation of an object's velocity as a function of time. It's a fundamental tool in physics to analyse and visualize motion.

Key Features:

1. Time (t) is plotted on the x-axis.

2. Velocity (v) is plotted on the y-axis.

3. The graph shows the object's velocity at various times.

Types of Motion on a Velocity-Time Graph:

1. Horizontal line: Uniform motion (constant velocity).

2. Straight line with slope: Uniformly accelerated motion (constant acceleration).

3. Curved line: Non-uniformly accelerated motion.

4. zig-zag or oscillating line: Periodic motion (vibration or oscillation).

Importance:

1. Analyses acceleration: Slope represents acceleration.

2. Determines velocity: y-axis value at a given time.

3. Calculates displacement: Area under the curve represents displacement.

4. Identifies patterns: Periodic, oscillatory, or chaotic behaviour.

5. Compares motions: Different objects or scenarios.

6. Solves problems: Finds velocity, acceleration, and displacement.

7. Real-world applications: Trajectory planning, motion control systems, physics simulations.

Interpretation:

1. Slope ($\Delta v/\Delta t$) represents acceleration.

2. Area under the curve represents displacement.

3. Intersection with x-axis indicates instant rest (zero velocity).

4. Asymptotes indicate constant velocity.

Real-World Examples:

1. Vehicle motion: Acceleration, braking, and cruising.

2. Projectile motion: Trajectory of a thrown ball or launched rocket.

3. Sports analytics: Athlete's velocity and acceleration.

4. Mechanical systems: Motor performance and gearbox optimization.

Advantages:

1. Visualizes complex motions.

2. Facilitates calculations.

3. Helps predict future motion.

4. Enhances understanding of physical phenomena.

Limitations:

1. Does not show position directly.

2. Requires additional calculations for displacement.

By analysing velocity-time graphs, you can gain insights into various physical phenomena, making it an essential tool for physicists, engineers, and scientists.

Common velocity-time graph shapes and their interpretations:

1. Triangle: Uniformly accelerated motion.

2. Trapezoid: Uniform motion with initial and final acceleration.

3. S-curve: Non-uniformly accelerated motion.

4. Sinusoidal: Periodic motion.

Vectors to Velocity: A Guide to Motion and Mechanics *By: Prashant Kumar Lal*

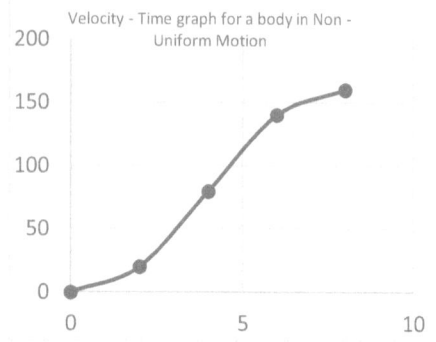

Time (s)	0	2	4	6	8
Velocity (m/s)	0	20	80	140	160

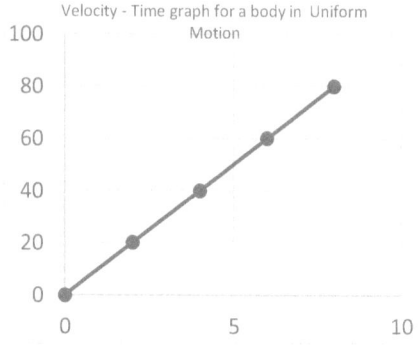

Time (s)	0	2	4	6	8
Velocity (m/s)	0	20	40	60	80

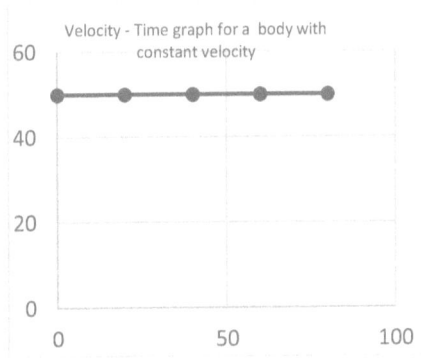

Time (s)	0	20	40	60	80
Velocity (m/s)	50	50	50	50	50

- **Acceleration-Time Graph: Shows acceleration vs. time**.

An acceleration-time graph is a graphical representation of an object's acceleration as a function of time. It's a fundamental tool in physics to analyse and visualize motion.

Key Features:

1. Time (t) is plotted on the x-axis.

2. Acceleration (a) is plotted on the y-axis.

3. The graph shows the object's acceleration at various times.

Types of Motion on an Acceleration-Time Graph:

1. Horizontal line: Constant acceleration.

2. Straight line with slope: Changing acceleration (jerk).

3. Curved line: Non-uniformly changing acceleration.

4. zig-zag or oscillating line: Periodic acceleration.

Importance:

1. Analyses jerk: Rate of change of acceleration.

2. Determines acceleration: y-axis value at a given time.

3. Calculates change in velocity: Area under the curve represents Δv.

4. Identifies patterns: Periodic, oscillatory, or chaotic behaviour.

5. Compares motions: Different objects or scenarios.

6. Solves problems: Finds acceleration, velocity, and displacement.

7. Real-world applications: Vehicle safety, robotics, aerospace engineering.

Interpretation:

1. Slope ($\Delta a / \Delta t$) represents jerk.

2. Area under the curve represents change in velocity (Δv).

3. Intersection with x-axis indicates no acceleration.

4. Asymptotes indicate constant acceleration.

Real-World Examples:

1. Vehicle safety: Braking and crash testing.

2. Aerospace engineering: Rocket propulsion and re-entry.

3. Robotics: Motion control and navigation.

4. Sports analytics: Athlete's acceleration and performance.

Advantages:

1. Visualizes complex motions.

2. Facilitates calculations.

3. Helps predict future motion.

4. Enhances understanding of physical phenomena.

Limitations:

1. Does not show velocity or position directly.

2. Requires additional calculations for velocity and displacement.

Common acceleration-time graph shapes:

1. Rectangle: Constant acceleration.

2. Triangle: Linearly changing acceleration.

3. S-curve: Non-linearly changing acceleration.

4. Sinusoidal: Periodic acceleration.

Importance in various fields:

1. Physics and Engineering: Analyses motion, force, and energy.

2. Computer Science: Simulates realistic motion in games and animations.

3. Biomechanics: Studies human movement and athletic performance.

4. Aerospace Engineering: Designs efficient propulsion systems.

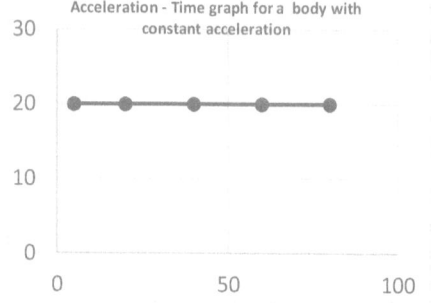

Time (s)	5	20	40	60	80
Acceleration in m/s²	20	20	20	20	20

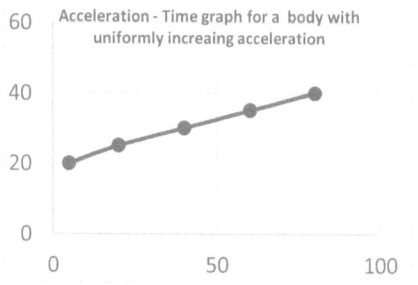

Time (s)	5	20	40	60	80
Acceleration in m/s²	20	25	30	35	40

Vectors to Velocity: A Guide to Motion and Mechanics   *By: Prashant Kumar Lal*

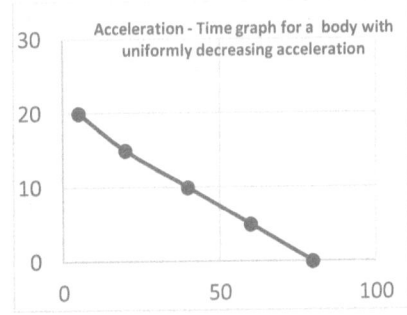

Time (s)	5	20	40	60	80
Acceleration in m/s^2	20	15	10	5	0

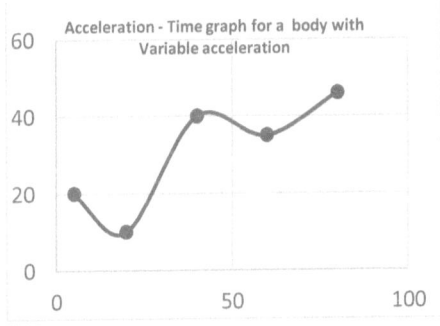

Time (s)	5	20	40	60	80
Acceleration in m/s^2	20	10	40	35	46

To prove graphically that v = u + at, we can use a velocity-time graph.

Step 1: Draw the velocity-time graph

Draw a graph with time (t) on the x-axis and velocity (v) on the y-axis.

Step 2: Plot initial velocity (u)

Plot a point on the y-axis representing the initial velocity (u).

Step 3: Plot acceleration (a)

To draw a line with slope 'a' (acceleration) starting from the initial velocity point, we'll use the concept of slope-intercept form:

$y = mx + b$

where:

- y is the final velocity

- m is the slope (acceleration 'a')

- x is the time

- b is the initial velocity

Let's assume:

- Initial velocity (v_0) = 5 m/s

- Acceleration (a) = 2 m/s²

- Time (t) = 4 s

Now, we can plug in the values:

$v = v_0 + at$

$v = 5 + 2t$

To draw the line, we'll find two points:

Point 1 (t = 0):

$v = 5 + 2(0) = 5$

Point 2 (t = 4):

$v = 5 + 2(4) = 13$

Now, we can draw the line:

Let

Time (s)	0	1	2	3	4
Velocity (m/s)	5	10	15	20	25

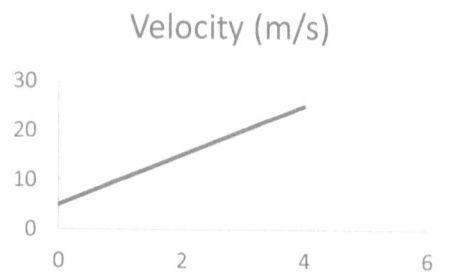

The line starts at (0, 5) with a slope of 2, representing an acceleration of 2 m/s².

Please note that this is a simple representation. For more complex scenarios, you may need to use graphical software or calculators.

Step 4: Add time (t)

Choose a point on the x-axis representing time 't'.

Step 5: Find final velocity (v)

Draw a vertical line from the chosen time point to intersect the acceleration line. The y-coordinate of this intersection point represents the final velocity (v).

Observation

Measure the vertical distance between the initial velocity point and the final velocity point. This distance represents the change in velocity (v - u).

Calculation

Calculate the area under the accceleration line between t = 0 and t = t. This area represents the product of acceleration and time (at).

Result

The vertical distance (v - u) is equal to the area under the acceleration line (at).

Conclusion

Graphically, we have shown that:

v - u = at

Rearranging the equation:

v = u + at

Thus, we have graphically proven the equation v = u + at.

This graphical representation illustrates how the velocity changes over time under constant acceleration.

To prove graphically that S = ut + 1/2at^2, we can use a velocity-time graph and a displacement-time graph.

Step 1: Draw the velocity-time graph

Draw a graph with time (t) on the x-axis and velocity (v) on the y-axis.

Step 2: Plot initial velocity (u)

Plot a point on the y-axis representing the initial velocity (u).

Step 3: Plot acceleration (a)

Draw a line with slope 'a' (acceleration) starting from the initial velocity point.

Step 4: Add time (t)

Choose a point on the x-axis representing time 't'.

Step 5: Find displacement (S)

Draw a vertical line from the chosen time point to intersect the velocity line. The area under the velocity line between t = 0 and t = t represents the displacement (S).

Calculation

1. Area under the velocity line = Area of rectangle + Area of triangle

2. Area of rectangle = u × t

3. Area of triangle = 1/2 × (at) × t

4. Total area (displacement) = ut + $1/2 at^2$

Result

The area under the velocity line, representing displacement (S), is equal to $ut + 1/2at^2$.

Conclusion

Graphically, we have shown that:

$S = ut + 1/2at^2$

Thus, we have graphically proven the equation $S = ut + 1/2at^2$.

This graphical representation illustrates how displacement changes over time under constant acceleration.

Key Points:

- The rectangle area represents the displacement due to initial velocity (ut).

- The triangle area represents the additional displacement due to acceleration ($1/2at^2$).

- The total area represents the total displacement (S).

To prove graphically that $v^2 = u^2 + 2as$, we'll use:

1. Velocity-time graph

2. Displacement-time graph

3. Graphical calculation

Velocity-Time Graph:

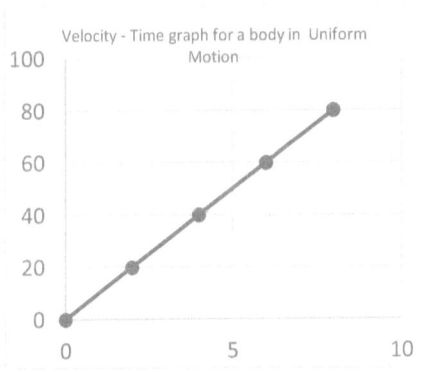

Graphical Calculation:

1. Area under velocity-time graph = Displacement (s)

2. Area = $ut + 1/2at^2$ (from $S = ut + 1/2at^2$)

3. $v^2 = (u + at)^2 = u^2 + 2uat + a^2t^2$

4. $v^2 = u^2 + 2a(ut + 1/2at^2)$

(substitute $s = ut + 1/2at^2$)

5. $v^2 = u^2 + 2as$

Displacement-time graph

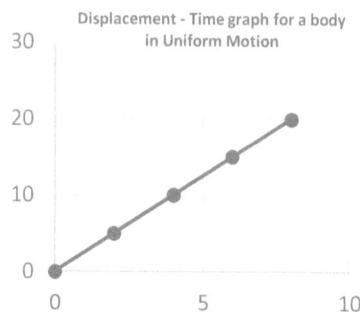

Graphical Representation:

Triangle ABC:

A (0, u)

B (t, v)

C (t, u)

Area of Triangle ABC:

= 1/2 × base × height

$= 1/2 \times t \times (v - u)$

$= 1/2 \times t \times at$

$= 1/2at^2$

Area of Rectangle ADEB:

$= $ base $\times$ height

$= t \times u$

$= ut$

Total Area (Displacement):

$= ut + 1/2at^2$

$= s$

Graphical Proof:

$V^2 = u^2 + 2as$

This graphical proof demonstrates the relationship between velocity, initial velocity, acceleration, and displacement.

Key Points:

- The area under the velocity-time graph represents displacement.

- The rectangle and triangle areas represent the displacement components.

- The graphical calculation shows the equivalence of v^2 and $u^2 + 2as$.

1.6 Relative Motion

Relative Motion in Kinematics: Understanding Motion from Different Perspectives

Introduction

Imagine you're on a train, and you throw a ball straight up in the air. To you, on the train, the ball goes straight up and comes straight back down. But what about someone watching from outside the train? They'll see the ball moving in a curved path! This difference in perspective is what relative motion is all about.

What is Relative Motion?

Relative motion is the study of how objects move in relation to each other. It's essential to understand that motion is not always absolute; it depends on the observer's reference frame.

Key Concepts

1. Reference Frame: A coordinate system used to describe motion.

2. Relative Motion: Motion between objects in different reference frames.

3. Observer: A person or object observing motion from a specific reference frame.

4. Frame of Reference: Inertial (stationary or moving at constant velocity) or non-inertial (accelerating).

Types of Relative Motion

1. **Translational Relative Motion: Objects move along parallel paths**.

Introduction

Equations

1. **Relative Velocity**: $V_{rel} = V_A - V_B$ (relative velocity between objects A and B)

2. **Relative Position:** $r_{rel} = r_A - r_B$ (relative position between objects A and B)

Examples

1. Car Passing Another Car: Relative motion between two cars.

2. Swimmer in a River: Relative motion between swimmer and riverbank.

3. Two Trains Moving on Parallel Tracks: Relative motion between trains.

Solving Problems

1. Identify Reference Frames: Determine the observer's frame of reference.

2. Use Relative Velocity Equation: Apply $V_{rel} = V_A - V_B$.

3. Visualize Motion: Draw diagrams to understand relative motion.

Real-World Applications

1. Traffic Flow: Understanding relative motion helps optimize traffic flow.

2. Air Traffic Control: Relative motion ensures safe distance between aircraft.

3. Robotics: Relative motion enables robots to navigate and interact.

Common Mistakes

1. Forgetting to subtract velocities: Remember to use $V_{rel} = V_A - V_B$

2. Ignoring reference frames: Ensure you're using the correct reference frame.

Practice Questions

1. Two cars travel in the same direction. Car A is at 60 km/h, Car B at 80 km/h. What is their relative velocity?

2. A swimmer swims at 5 km/h in a river flowing at 3 km/h. What is their relative velocity?

3. Two trains travel on parallel tracks. Train A is at 100 km/h, Train B at 120 km/h. What is their relative velocity?

Glossary

1. Relative velocity: Velocity between objects in different reference frames.

2. Reference frame: Coordinate system used to describe motion.

3. Parallel paths: Paths that never intersect.

By mastering translational relative motion, you'll better understand how objects move in relation to each other, solving problems in physics, engineering, and real-world applications.

Imagine two cars driving on parallel roads. One car is traveling at 60 km/h, while the other is traveling at 80 km/h. How do we describe their motion relative to each other? This is where translational relative motion comes in.

What is Translational Relative Motion?

Translational relative motion occurs when two or more objects move along parallel paths, with their motion described relative to each other.

Key Concepts

1. Parallel Paths: Objects move along paths that never intersect.

2. Relative Velocity: Velocity of one object with respect to another.

3. Reference Frame: Coordinate system used to describe motion.

Types of Translational Relative Motion

   a. Same Direction: Objects move in the same direction.
   b. Opposite Direction: Objects move in opposite directions.
   c. Relative Rest: Objects are at rest relative to each other.
2. **Rotational Relative Motion: Objects rotate relative to each other**.

Introduction

Imagine two gears connected to each other. As one gear rotates clockwise, the other gear rotates counterclockwise. How do we describe their rotational motion relative to each other? This is where rotational relative motion comes in.

What is Rotational Relative Motion?

Rotational relative motion occurs when two or more objects rotate relative to each other, with their motion described in terms of angular displacement, velocity, and acceleration.

Key Concepts

1. Angular Displacement: Measure of rotation (θ)

2. Angular Velocity: Rate of rotation (ω)

3. Angular Acceleration: Rate of change of angular velocity (α)

4. Reference Frame: Coordinate system used to describe rotation

Types of Rotational Relative Motion

1. Same Axis Rotation: Objects rotate around the same axis.

2. Different Axis Rotation: Objects rotate around different axes.

3. Relative Rotation: Rotation of one object relative to another.

Equations

1. Relative Angular Velocity: $\omega_{rel} = \omega_A - \omega_B$ (relative angular velocity)

2. Relative Angular Displacement: $\theta_{rel} = \theta_A - \theta_B$ (relative angular displacement)

3. Relative Angular Acceleration: $\alpha_{rel} = \alpha_A - \alpha_B$ (relative angular acceleration)

Examples

1. Gears and Pulleys: Rotational motion in mechanical systems.

2. Earth's Rotation and Revolution: Relative rotation between Earth, Sun, and Moon.

3. Bicycle Wheels: Relative rotation between wheels and pedals.

Solving Problems

1. Identify Reference Frames: Determine the observer's frame of reference.

2. Use Relative Angular Velocity Equation: Apply $\omega_{rel} = \omega_A - \omega_B$.

3. Visualize Motion: Draw diagrams to understand rotational relative motion.

Real-World Applications

1. Robotics: Relative rotation enables robots to grasp and manipulate objects.

2. Aerospace Engineering: Relative rotation crucial for spacecraft navigation.

3. Mechanical Engineering: Relative rotation essential for gear design.

Common Mistakes

1. Forgetting to subtract angular velocities: Remember to use

$\omega_{rel} = \omega_A - \omega_B$.

2. Ignoring reference frames: Ensure you're using the correct reference frame.

Practice Questions

1. Two gears rotate with angular velocities 100 rpm and 50 rpm. What is their relative angular velocity?

2. A bicycle wheel rotates at 10 rad/s. The pedal rotates at 5 rad/s. What is their relative angular velocity?

3. Earth rotates at 360°/day. The Moon orbits Earth at 360°/28 days. What is their relative angular velocity?

Glossary

1. Angular displacement: Measure of rotation (θ).

2. Angular velocity: Rate of rotation (ω).

3. Angular acceleration: Rate of change of angular velocity (α).

By mastering rotational relative motion, you'll better understand how objects rotate and interact, solving problems in physics, engineering, and real-world applications.

3. Combined Motion: Combination of translational and rotational motion.

Introduction

Imagine a rolling wheel, a flying helicopter, or a spinning top. These objects exhibit both translational (linear) and rotational motion simultaneously. This combination of motions is called combined motion.

What is Combined Motion?

Combined motion occurs when an object experiences both translational (linear) and rotational motion at the same time.

Key Concepts

1. Translational Motion: Linear motion (movement along a straight line)

2. Rotational Motion: Rotation around a fixed axis

3. Combined Motion: Combination of translational and rotational motion

Types of Combined Motion

1. Rectilinear Rotation: Rotation around a fixed axis while moving in a straight line

2. Curvilinear Rotation: Rotation around a fixed axis while moving along a curved path

3. Helical Motion: Rotation around a fixed axis while moving along a helical path

Examples

1. Rolling Wheel: Translational motion along the road and rotational motion around the axle

2. Flying Helicopter: Translational motion through the air and rotational motion of the blades

3. Spinning Top: Translational motion along the surface and rotational motion around its axis

Equations

1. Translational Velocity: $v = \Delta x/\Delta t$

2. Rotational Velocity:

$\omega = \Delta\theta/\Delta t$

3. Combined Motion Equation: $r(t) = r_{trans}(t) + r_{rot}(t)$

Solving Problems

1. **Identify Translational and Rotational Components:** Separate motion into translational and rotational parts

2. **Use Combined Motion Equation:** Apply
$r(t) = r_{trans}(t) + r_{rot}(t)$

3. Visualize Motion: Draw diagrams to understand combined motion

Relative Motion Equations

1. **Velocity Addition:**

$V_{rel} = V_A + V_B$ (relative velocity between objects A and B)

2. Position Vector: $R_{rel} = R_A - R_B$ (relative position vector between objects A and B)

3. Acceleration: $a_{rel} = a_A - a_B$ (relative acceleration between objects A and B)

Real-Life Examples

1. Car passing another car: Relative motion between two cars.

2. Throwing a ball from a moving train: Relative motion between ball, train, and observer.

3. Earth's rotation and revolution: Relative motion between Earth, Sun, and observer.

Why is Relative Motion Important?

1. Simplifies complex problems: Breaks down motion into manageable components.

2. Accurate predictions: Essential for navigation, transportation, and astronomy.

3. Real-world applications: GPS, aircraft navigation, and robotics.

Common Misconceptions

1. Assuming absolute motion: Motion is relative, not absolute.

2. Ignoring reference frames: Different observers can see different motions.

Tips for Problem-Solving

1. Identify reference frames: Determine the observer's frame of reference.

2. Use relative motion equations: Apply velocity addition, position vector, and acceleration equations.

3. Visualize motion: Draw diagrams to understand relative motion.

Conclusion

Relative motion in kinematics helps us understand how motion is perceived from different perspectives. By mastering relative motion, you'll develop problem-solving skills, critical thinking, and a deeper understanding of the physical world.

Practice Questions

1. A car travels at 60 km/h. A bicycle travels at 20 km/h in the same direction. What is their relative velocity?

2. An airplane flies at 500 km/h. A passenger walks towards the cockpit at 5 km/h. What is their relative velocity?

3. A ball is thrown from a moving train. Describe the motion from the train passenger's and outside observer's perspectives.

Glossary

1. Inertial frame: A reference frame moving at constant velocity.

2. Non-inertial frame: A reference frame accelerating.

3. Relative velocity: Velocity between objects in different reference frames.

Here are twenty short answer questions based on Frame of Reference, Concept of point and extended objects, and Types of frames (inertial and non-inertial:

Frame of Reference (1-5)

1. What is a frame of reference?

Answer: A coordinate system used to describe motion.

2. What are the essential components of a frame of reference?

Answer: Origin, axes (x, y, z), and unit vectors.

3. Why is a frame of reference necessary?

Answer: To describe motion accurately and unambiguously.

4. What is the difference between an inertial and non-inertial frame?

Answer: Inertial frames move at constant velocity, while non-inertial frames accelerate.

5. Give an example of a frame of reference.

Answer: A car's motion described relative to the road.

Concept of Point and Extended Objects (6-10)

6. Define a point object.

Answer: An object with negligible size.

7. What is an extended object?

Answer: An object with finite size.

8. How do point objects move?

Answer: Through translation and rotation.

9. What is the centre of mass of an extended object?

Answer: The average position of the object's mass.

10. Give an example of an extended object.

Answer: A rotating wheel.

Types of Frames (Inertial and Non-Inertial) (11-20)

11. Define an inertial frame of reference.

Answer: A frame moving at constant velocity.

12. What is a non-inertial frame of reference?

Answer: A frame accelerating or rotating.

13. Give an example of an inertial frame.

Answer: A car moving at constant speed.

14. What is the advantage of using inertial frames?

Answer: Simplifies motion description.

15. Can a rotating frame be inertial?

Answer: No.

16. What is the difference between inertial and non-inertial forces?

Answer: Inertial forces are real, while non-inertial forces are fictitious.

17. Give an example of a non-inertial frame.

Answer: A merry-go-round.

18. How does time dilation relate to inertial frames?

Answer: Time dilation occurs in non-inertial frames.

19. Can an accelerated frame be inertial?

Answer: No.

20. What is the significance of choosing the right frame of reference?

Answer: Accurate motion description and analysis.

Here are twenty short answer questions based on Frame of Reference, Concept of point and extended objects, and Types of frames (inertial and non-inertial:

Frame of Reference (1-5)

1. What is a frame of reference in physics?

Answer: A coordinate system used to describe motion.

2. What are the key components of a frame of reference?

Answer: Origin, axes (x, y, z), and unit vectors.

3. Why is a frame of reference necessary in physics?

Answer: To describe motion accurately and unambiguously.

4. What is the difference between an observer and a frame of reference?

Answer: An observer uses a frame of reference to measure motion.

5. Give an example of a frame of reference.

Answer: A car's motion described relative to the road.

Concept of Point and Extended Objects (6-10)

6. Define a point object in physics.

Answer: An object with negligible size.

7. What is an extended object in physics?

Answer: An object with finite size.

8. How do point objects move?

Answer: Through translation and rotation.

9 What is the centre of mass of an extended object?

Answer: The average position of the object's mass.

10. Give an example of an extended object.

Answer: A rotating wheel.

Types of Frames (Inertial and Non-Inertial) (11-20)

11. Define an inertial frame of reference.

Answer: A frame moving at constant velocity.

12. What is a non-inertial frame of reference?

Answer: A frame accelerating or rotating.

13. Give an example of an inertial frame.

Answer: A car moving at constant speed.

14. What is the advantage of using inertial frames?

Answer: Simplifies motion description.

15. Can a rotating frame be inertial?

Answer: No.

16. What is the difference between inertial and non-inertial forces?

Answer: Inertial forces are real, while non-inertial forces are fictitious.

17. Give an example of a non-inertial frame.

Answer: A merry-go-round.

18. How does time dilation relate to inertial frames?

Answer: Time dilation occurs in non-inertial frames.

19. Can an accelerated frame be inertial?

Answer: No.

20. What is the significance of choosing the right frame of reference?

Answer: Accurate motion description and analysis.

Here are some NCERT-style questions based on the given topics:

NCERT Questions

1. Give an example of a non-inertial frame.

(**Answer**: A merry-go-round)

NCERT Question 1

A child is sitting on a merry-go-round that is rotating at a constant rate. What type of frame of reference is this?

Solution

This is a non-inertial frame of reference because the merry-go-round is rotating, which means it is accelerating.

2. How does time dilation relate to inertial frames?

(**Answer**: Time dilation occurs in non-inertial frames)

NCERT Question 2

A spaceship is moving at high-speed relative to an observer on Earth. Which of the following statements is true?

A) Time dilation occurs in the inertial frame.

B) Time dilation occurs in the non-inertial frame.

C) Time dilation does not occur.

Solution

B) Time dilation occurs in the non-inertial frame.

3. Can an accelerated frame be inertial?

(**Answer**: No)

NCERT Question 3

A car is moving along a straight road with increasing speed. Is this an inertial or non-inertial frame of reference?

Solution

This is a non-inertial frame of reference because the car is accelerating.

4. What is the significance of choosing the right frame of reference?

(**Answer:** Accurate motion description and analysis)

NCERT Question 4

Why is it essential to choose the correct frame of reference when describing motion?

Solution

Choosing the correct frame of reference is crucial because it allows for accurate description and analysis of motion. The wrong frame of reference can lead to incorrect conclusions.

Additional NCERT Questions

5. A particle is moving in a circular path. What type of frame of reference is suitable for describing its motion?

Solution

A non-inertial frame of reference.

6. What is the difference between inertial **and non-inertial forces?**

Solution

Inertial forces are real forces, while non-inertial forces are fictitious forces that arise due to acceleration.

7. A person is standing on a train moving at constant velocity. Is this an inertial or non-inertial frame of reference?

Solution

Inertial frame of reference.

8. Give an example of an inertial frame of reference.

Solution

A car moving at constant speed.

9. What is the centre of mass of an extended object?

Solution

The average position of the object's mass.

10. Define a point object.

Solution

An object with negligible size.

Here are twenty numerical problems based on Frame of Reference, Concept of point and extended objects, and Types of frames (inertial and non-inertial), categorized as difficult and most difficult:

Difficult (1-10)

1. A particle moves with an initial velocity of 20 m/s and accelerates at 5 m/s² for 3 s. Find its final velocity and displacement.

Solution:

u = 20 m/s

a = 5 m/s²

t = 3 s

v = u + at

= 20 + 5(3)

= 20 + 15

= 35 m/s

$s = ut + 1/2at^2$

$= 20(3) + 1/2(5)(3)^2$

= 60 + 22.5

= 82.5 m

2. A wheel rotates at 10 rad/s. Find its linear velocity at a distance of 4 m.

ω = 10 rad/s

r = 4 m

v = ωr

= 10(4)

= 40 m/s

3. A car travels 200 m in 10 s. Find its average velocity and acceleration.

s = 200 m

t = 10 s

v_{avg} = s/t

= 200/10

$= 20$ m/s

$a = \Delta v/\Delta t$

$= (20 - 0)/10$

$= 2$ m/s²

4. A particle moves in a circular path with a radius of 5 m and angular velocity of 2 rad/s. Find its linear velocity.

$\omega = 2$ rad/s

$r = 5$ m

$v = \omega r$

$- 2(5)$

$= 10$ m/s

5. A train travels from city A to city B at 80 km/h. Find its displacement in 2 hours.

$v = 80$ km/h

$t = 2$ h

$s = vt$

= 80(2)

= 160 km

6. A cyclist moves 300 m in 15 s. Find its average velocity and acceleration.

s = 300 m

t = 15 s

v_{avg} = s/t

= 300/15

= 20 m/s

a = $\Delta v/\Delta t$

= (20 - 0)/15

= 1.33 m/s²

7. A wheel rotates 5 times in 2 s. Find its angular velocity.

θ = 5(2π)

t = 2 s

ω = θ/t

= 10π/2

$= 5\pi$ rad/s

8. A particle moves with an initial velocity of 15 m/s and accelerates at 3 m/s² for 2 s. Find its final velocity.

u = 15 m/s

a = 3 m/s²

t = 2 s

v = u + at

= 15 + 3(2)

= 15 + 6

= 21 m/s

9. A car travels 150 m in 5 s. Find its average velocity and acceleration.

s = 150 m

t = 5 s

$v_{avg} = s/t$

= 150/5

= 30 m/s

$a = \Delta v/\Delta t$

$= (30 - 0)/5$

$= 6$ m/s²

10. A train travels at 100 km/h. Find its displacement in 1.5 hours.

$v = 100$ km/h

$t = 1.5$ h

$s = v\,t$

$= 100(1.5)$

$= 150$ km

Most Difficult (11-13)

11. A particle moves in a circular path with a radius of 10 m and angular velocity of 3 rad/s. Find its linear velocity and acceleration.

$\omega = 3$ rad/s

$r = 10$ m

$v = \omega r$

$= 3(10)$

$= 30$ m/s

$a = \omega^2 r$

$= 3^2 (10)$

$= 90$ m/s²

12. A car travels 400 m in 10 s. Find its average velocity, acceleration, and angular velocity.

$s = 400$ m

$t = 10$ s

$v_{avg} = s/t$

$= 400/10$

$= 40$ m/s

$a = \Delta v / \Delta t$

$= (40 - 0)/10$

$= 4$ m/s²

$\omega = v/r$

$= 40/10$

= 4 rad/s

13. A wheel rotates 10 times in 4 s. Find its angular velocity and linear velocity.

$\theta = 10(2\pi)$

t = 4 s

$\omega = \theta/t$

$= 20\pi/4$

$= 5\pi$ rad/s

v = ω r

$= 5\pi (5)$

$= 25\pi$ m/s

Here's an exercise of questions based on Frame of Reference, Concept of point and extended objects, and Types of frames (inertial and non-inertial):

Exercise

Section A: Multiple Choice Questions (1-5)

1. What is the primary purpose of a frame of reference?

A) To describe motion

B) To measure time

C) To calculate force

D) To determine energy

Hint: Think about the definition of a frame of reference.

2. Which of the following is an example of a point object?

A) A car

B) A ball

C) A building

D) A tree

Hint: *Consider the size and dimensions of each option.*

3. What type of frame is accelerating?

A) Inertial

B) Non-inertial

C) Stationary

D) Rotating

Hint: *Recall the definition of inertial and non-inertial frames.*

4. A particle moves in a circular path. What type of motion is this?

A) Translational

B) Rotational

C) Circular

D) Linear

Hint: *Think about the path of the particle.*

5. Which frame is suitable for describing motion on Earth?

A) Inertial

B) Non-inertial

C) Geocentric

D) Heliocentric

Hint: Consider Earth's rotation and revolution.

Section B: Short Answer Questions (6-12)

6. Define a frame of reference. (3 marks)

Hint: *Describe the components of a frame of reference.*

7. What is the difference between inertial and non-inertial frames? (4 marks)

Hint: *Explain the characteristics of each type.*

8. Describe the concept of point objects. (3 marks)

Hint: *Discuss size and dimensions.*

9. A car travels 100 m in 10 s. Find its average velocity. (4 marks)

Hint: *Use $v = s/t$.*

10. What is the centre of mass of an extended object? (3 marks)

Hint: *Define centre of mass.*

11. A wheel rotates 5 times in 2 s. Find its angular velocity. (4 marks)

Hint: *Use $\omega = \theta/t$.*

12. Describe the concept of extended objects. (3 marks)

Hint: *Discuss size and dimensions.*

Section C: Long Answer Questions (13-17)

13. Derive the equation $v = u + at$. (6 marks)

Hint: *Use kinematic equations.*

14. Explain the concept of relative motion. (6 marks)

Hint: *Discuss frames of reference.*

15. Describe the differences between translational and rotational motion. (6 marks)

Hint: *Explain each type.*

16. A particle moves in a circular path with radius 5 m and angular velocity 2 rad/s. Find its linear velocity. (6 marks)

Hint: *Use $v = \omega r$.*

17. Explain the significance of choosing the right frame of reference. (6 marks)

Hint: *Discuss accuracy.*

Section D: Numerical Problems (18-21)

18. A car travels 200 m in 10 s. Find its average velocity and acceleration.

Hint: Use $v = s/t$ *and* $a = \Delta v/\Delta t$.

19. A wheel rotates 10 times in 4 s. Find its angular velocity.

Hint: *Use $\omega = \theta/t$.*

20. A particle moves with initial velocity 20 m/s and accelerates at 5 m/s² for 3 s. Find its final velocity.

Hint: Use $v = u + at$.

21. A train travels 500 m in 20 s. Find its average velocity.

Hint: *Use $v = s/t$.*

22. A cyclist moves 300 m in 15 s. Find its average velocity and acceleration.

Hint: *Use $v = s/t$ and $a = \Delta v/\Delta t$.*

⁕⋆⋆⋆⋆⁕

Vectors to Velocity: A Guide to Motion and Mechanics *By: Prashant Kumar Lal*

Chapter – 2

Motion in a Plane

Scalar and vector quantities, position and displacement vectors

Equality of vectors, multiplication of vectors by a real number; addition and subtraction of vectors

Relative velocity

Motion in a plane, cases of uniform velocity and uniform acceleration, projectile motion

Uniform circular motion

Motion in a Plane

Introduction

In physics, motion is the change in position of an object with respect to time. So far, we've explored motion in one dimension, but in real life, objects can move in multiple directions. In this chapter, we'll delve into motion in a plane, where objects move in two dimensions.

What is Motion in a Plane?

Motion in a plane occurs when an object moves in two dimensions, typically represented by the x-y plane. The object's position is described by its x and y coordinates, which change over time.

Types of Motion in a Plane

There are several types of motion in a plane:

1. **Rectilinear Motion**: The object moves in a straight line, either horizontally or vertically.

2. **Circular Motion**: The object moves in a circular path, with its

distance from the centre remaining constant.

3. **Curvilinear Motion**: The object moves along a curved path, which can be a combination of circular and rectilinear motion.

Describing Motion in a Plane

To describe motion in a plane, we use the following parameters:

1. **Position**: The object's location in the x-y plane, represented by its x and y coordinates.
2. **Displacement**: The change in position, represented by the vector $\Delta r = (\Delta x, \Delta y)$.
3. **Velocity**: The rate of change of position, represented by the vector $v = (v_x, v_y)$.
4. **Acceleration**: The rate of change of velocity, represented by the vector $a = (a_x, a_y)$.

Graphical Representation

Motion in a plane can be represented graphically using:

1. **Position-Time Graphs**: Show the object's position as a function of time.
2. **Velocity-Time Graphs**: Show the object's velocity as a function of time.
3. **Acceleration-Time Graphs**: Show the object's acceleration as a function of time.

Equations of Motion in a Plane

The equations of motion in a plane are:

1. Position: $r(t) = (x(t), y(t))$
2. Velocity: $v(t) = (v_x(t), v_y(t))$
3. Acceleration: $a(t) = (a_x(t), a_y(t))$

Solving Problems in Motion in a Plane

To solve problems in motion in a plane, follow these steps:

1. **Identify the type of motion**: Determine if the motion is rectilinear, circular, or curvilinear.

2. **Choose a coordinate system**: Select a suitable coordinate system to describe the motion.

3. **Write the equations of motion**: Use the equations of motion to describe the object's position, velocity, and acceleration.

4. **Solve the equations**: Use algebraic and graphical methods to solve the equations.

Examples and Numerical Problems

1. A particle moves in a circular path with a radius of 5 m. Find its velocity and acceleration.

Answer: $v = 2\pi r/T$, $a = v^2/r$

2. An object moves in a straight line with a velocity of 5 m/s. Find its position after 10 s.

Answer: $x(t) = v_x(t)$, $y(t) = 0$

Questions with Hints
Multiple Choice Questions

1. What is the type of motion that occurs when an object moves in a straight line?

A) Rectilinear

B) Circular

C) Curvilinear

D) Rotational

Hint: *Think about the path of the object.*

Answer: A) Rectilinear

2. What is the unit of velocity in the x-y plane?

A) m/s

B) km/h

C) mph

D) rad/s

Hint: *Recall velocity units.*

Answer: A) m/s

Short Answer Questions

1. Describe the difference between rectilinear and circular motion.

Hint: *Explain the path of the object.*

Answer: Rectilinear motion occurs in a straight line, while circular motion occurs in a circular path.

2. Calculate the velocity of an object moving in a circular path with a radius of 3 m and a period of 2 s.

Hint: *Use v = 2πr/T.*

Answer: v = 2π (3)/2 = 3π m/s

Long Answer Questions

1. Explain the concept of motion in a plane.

Hint: *Discuss the types of motion and the parameters used to describe motion.*

Answer: Motion in a plane occurs when an object moves in two dimensions, described by its x and y coordinates.

2. Calculate the acceleration of an object moving in a straight line with a velocity of 10 m/s and a displacement of 20 m.

Hint: *Use a = Δv/Δt.*

Answer: a = (10 - 0) / (20/10) = 5 m/s²

Scalar Quantities: A Comprehensive Guide

Introduction

In physics and engineering, we deal with various quantities that can be classified into two main categories: scalar and vector quantities. Scalar quantities are an essential part of mathematics and physics, and understanding them is crucial for problem-solving and critical thinking. In this article, we will explore the definition, examples, uses, and applications of scalar quantities, as well as comprehensive mathematical operations and dot products.

Definition of Scalar Quantities

A scalar quantity is a physical quantity that has only magnitude (amount or size) but no direction. Scalar quantities are described by a single number or value, which can be either positive or negative. Examples of scalar quantities include:

- Temperature
- Mass
- Time
- Distance
- Speed
- Energy

Examples of Scalar Quantities

1. The temperature of a room is 25°C, which is a scalar quantity.

2. The mass of a ball is 0.5 kg, which is a scalar quantity.

3. The time taken to complete a task is 2 hours, which is a scalar quantity.

4. The distance between two points is 5 meters, which is a scalar quantity.

5. The speed of a car is 60 km/h, which is a scalar quantity.

Uses and Applications of Scalar Quantities

Scalar quantities have numerous applications in various fields, including:

1. **Physics and engineering**: Scalar quantities are used to describe physical phenomena, such as temperature, mass, and energy.

2. **Mathematics**: Scalar quantities are used in algebra, geometry, and trigonometry.

3. **Computer science**: Scalar quantities are used in programming languages, such as C++ and Python.

4. **Data analysis**: Scalar quantities are used in data analysis and visualization.

Comprehensive Mathematical Operations

Scalar quantities can be manipulated using various mathematical operations, including:

1. Addition: The sum of two scalar quantities is a scalar quantity.

Example: 2 + 3 = 5

2. **Subtraction**: The difference between two scalar quantities is a scalar quantity.

Example: 5 - 2 = 3

3. **Multiplication**: The product of two scalar quantities is a scalar quantity.

Example: 4 × 5 = 20

4. **Division**: The quotient of two scalar quantities is a scalar quantity.

Example: 10 ÷ 2 = 5

Dot Products

A dot product is a mathematical operation that combines two scalar quantities to produce another scalar quantity. The dot product is denoted by a dot (·) and is calculated as follows:

$A \cdot B = |A| |B| \cos(\theta)$

where $|A|$ and $|B|$ are the magnitudes of the scalar quantities A and B, and θ is the angle between them.

Operation of Dot Products

The dot product has several properties and applications, including:

1. **Commutativity**:

$A \cdot B = B \cdot A$

2. **Distributivity**:

$A \cdot (B + C) = A \cdot B + A \cdot C$

3. **Scalar multiplication:**

$(kA) \cdot B = k(A \cdot B)$

Orthogonality: $A \cdot B = 0$ if A and B are orthogonal (perpendicular)

Examples of Dot Products

1. The dot product of two scalar quantities A = 3 and B = 4 is:

$A \cdot B = |A| |B| \cos(\theta) = 3 \times 4 \times \cos(0) = 12$

2. The dot product of two scalar quantities A = 2 and B = 5 is:

$A \cdot B = |A| |B| \cos(\theta) = 2 \times 5 \times \cos(\pi/2) = 0$

Applications of Dot Products

Dot products have numerous applications in physics, engineering, and computer science, including:

1. Calculating work and energy

2. Finding the angle between two vectors

3. Determining the orthogonality of two vectors

4. Solving systems of linear equations

Scalar quantities are an essential part of mathematics and physics, and understanding them is crucial for problem-solving and critical thinking. In this article, we have explored the definition, examples, uses, and applications of scalar quantities, as well as comprehensive mathematical operations and dot products. We hope that this guide has provided a comprehensive understanding of scalar quantities and their importance in various fields.

Vector Quantities: A Comprehensive Guide

Introduction

In physics and engineering, we deal with various quantities that can be classified into two main categories: *scalar and vector quantities*. Vector quantities are an essential part of mathematics and physics, and understanding them is crucial for problem-solving and critical thinking. In this article, we will explore the definition, examples, uses, and applications of vector quantities, as well as comprehensive mathematical operations and dot products.

Definition of Vector Quantities

A vector quantity is a physical quantity that has both magnitude (amount or size) and direction. Vector quantities are described by a pair of numbers or values, which represent the magnitude and direction of the quantity. Examples of vector quantities include:

- Displacement

- Velocity

- Acceleration

- Force

- Momentum

Examples of Vector Quantities

1. The displacement of an object from one point to another is 5 meters east, which is a vector quantity.

2. The velocity of a car is 60 km/h north, which is a vector quantity.

3. The acceleration of an object is 2 m/s² upward, which is a vector quantity.

4. The force applied to an object is 10 N west, which is a vector quantity.

5. The momentum of an object is 20 kg - m/s north, which is a vector quantity.

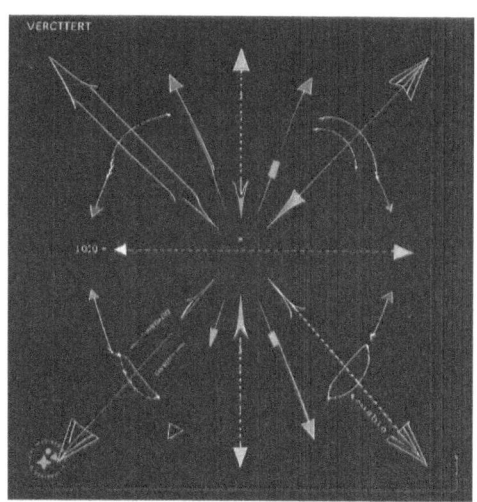

Uses and Applications of Vector Quantities

Vector quantities have numerous applications in various fields, including:

1. **Physics and engineering**: Vector quantities are used to describe physical phenomena, such as displacement, velocity, and force.

2. **Mathematics**: Vector quantities are used in geometry, trigonometry, and calculus.

3. **Computer science**: Vector quantities are used in programming languages, such as C++ and Python.

4. **Data analysis**: Vector quantities are used in data analysis and visualization.

Comprehensive Mathematical Operations

Vector quantities can be manipulated using various mathematical operations, including:

1. **Addition:** The sum of two vector quantities is a vector quantity.

Example: $A + B = (A_x + B_x, A_y + B_y)$

2. **Subtraction:** The difference between two vector quantities is a vector quantity.

Example: $A - B = (A_x - B_x, A_y - B_y)$

3. **Scalar multiplication**: The product of a scalar and a vector quantity is a vector quantity.

Example: $kA = (kA_x, kA_y)$

4. **Dot product**: The dot product of two vector quantities is a scalar quantity.

Example: $A \cdot B = |A| |B| \cos(\theta)$

Dot Products

A dot product is a mathematical operation that combines two vector quantities to produce a scalar quantity. The dot product is denoted by a dot ($\cdot$) and is calculated as follows:

$A \cdot B = |A| |B| \cos(\theta)$

where $|A|$ and $|B|$ are the magnitudes of the vector quantities A and B, and θ is the angle between them.

Operation of Dot Products

The dot product has several properties and applications, including:

1. **Commutativity**:
   $A \cdot B = B \cdot A$

2. **Distributivity**:

$A \cdot (B + C) = A \cdot B + A \cdot C$

3. Scalar multiplication:

$(kA) \cdot B = k(A \cdot B)$

4. **Orthogonality**: $A \cdot B = 0$ if A and B are orthogonal (perpendicular)

Examples of Dot Products

1. The dot product of two vector quantities $A = (3, 4)$ and $B = (2, 5)$ is:

$A \cdot B = |A| |B| \cos(\theta) = \sqrt{(3^2 + 4^2)} \times \sqrt{(2^2 + 5^2)} \times \cos(\theta) = 26$

2. The dot product of two vector quantities $A = (2, 3)$ and $B = (4, 5)$ is:

$A \cdot B = |A| |B| \cos(\theta) = \sqrt{(2^2 + 3^2)} \times \sqrt{(4^2 + 5^2)} \times \cos(\theta) = 23$

Applications of Dot Products

Dot products have numerous applications in physics, engineering, and computer science, including:

1. Calculating work and energy

2. Finding the angle between two vectors

3. Determining the orthogonality of two vectors

4. Solving systems of linear equations

Vector quantities are an essential part of mathematics and physics, and understanding them is crucial for problem-solving and critical thinking. In this article, we have explored the definition, examples, uses, and applications of vector quantities, as well as comprehensive mathematical operations and dot.

Vectors can be represented using arrows, with the direction of the arrow indicating the direction of the vector. Here's how to represent vectors using forward and backward arrows:

Forward Arrow (→)

- A forward arrow (→) is used to represent a vector in the positive direction.

- The arrow points in the direction of the vector.

- The length of the arrow represents the magnitude (length) of the vector.

Example: A vector with a magnitude of 5 units in the positive x-direction can be represented as:

5→

Backward Arrow (←)

- A backward arrow (←) is used to represent a vector in the negative direction.

- The arrow points in the opposite direction of the vector.

- The length of the arrow represents the magnitude (length) of the vector.

Example: A vector with a magnitude of 5 units in the negative x-direction can be represented as:

-5← or 5←

Notation

- Vectors can be represented using boldface letters (e.g., A, B, C) or with an arrow above the letter (e.g., $\vec{A}$, $\vec{B}$, $\vec{C}$).

- The direction of the vector can be indicated using a subscript (e.g., A_x, B_y, C_z).

Example:

- A = 3→ (a vector with a magnitude of 3 units in the positive x-direction)

- B = 4← (a vector with a magnitude of 4 units in the negative x-direction)

Addition and Subtraction of Vectors

- When adding vectors, the arrows are combined by connecting the tail of one arrow to the head of the other arrow.

- When subtracting vectors, the arrows are combined by connecting the tail of one arrow to the head of the other arrow, but with the direction of the second arrow reversed.

Example:

- A + B = (3→) + (4←) = 1→ (the resulting vector has a magnitude of 1 unit in the positive x-direction)

- A - B = (3→) - (4←) = 7→ (the resulting vector has a magnitude of 7 units in the positive x-direction)

Vectors: i^, j^, and k^
Introduction

In mathematics and physics, vectors are used to represent quantities with both magnitude and direction. In this article, we will explore the concept of i^, j^, and k^ in vectors, their mathematical operations, and applications.

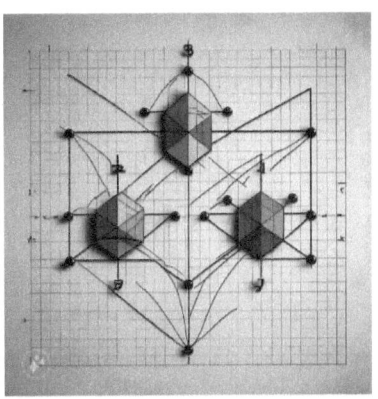

What are i^, j^, and k^?

i^, j^, and k^ are unit vectors that represent the x, y, and z axes in a three-dimensional coordinate system. They are used to describe the direction of a vector in space.

- i^ is the unit vector along the x-axis, representing the direction of the vector in the x-direction.

- $\hat{j}$ is the unit vector along the y-axis, representing the direction of the vector in the y-direction.

- $\hat{k}$ is the unit vector along the z-axis, representing the direction of the vector in the z-direction.

Mathematical Operations with $\hat{i}$, $\hat{j}$, and $\hat{k}$

1. **Addition**: Vectors can be added by combining their components.

Example: $A = 2\hat{i} + 3\hat{j}$, $B = 4\hat{i} + 5\hat{j}$

$A + B = (2 + 4)\hat{i} + (3 + 5)\hat{j} = 6\hat{i} + 8\hat{j}$

2. **Scalar Multiplication**: A vector can be multiplied by a scalar by multiplying each component.

Example: $A = 2\hat{i} + 3\hat{j}$, $k = 2$

$kA = 2(2\hat{i} + 3\hat{j}) = 4\hat{i} + 6\hat{j}$

3. Dot Product: The dot product of two vectors is a scalar quantity.

Example: $A = 2\hat{i} + 3\hat{j}$, $B = 4\hat{i} + 5\hat{j}$

$A \cdot B = (2)(4) + (3)(5) = 8 + 15 = 23$

4. Cross Product: The cross product of two vectors is a vector quantity.

Example: $A = 2\hat{i} + 3\hat{j}$, $B = 4\hat{i} + 5\hat{j}$

$A \times B = (2)(5) - (3)(4) = 10 - 12 = -2\hat{k}$

5. Magnitude: The magnitude of a vector is a scalar quantity.

Example: $A = 2\hat{i} + 3\hat{j}$

$|A| = \sqrt{(2^2 + 3^2)} = \sqrt{(4 + 9)} = \sqrt{13}$

6. Direction: The direction of a vector is a unit vector.

Example: $A = 2\hat{i} + 3\hat{j}$

Direction of $A = A / |A| = (2\hat{i} + 3\hat{j}) / \sqrt{13}$

Applications of i^, j^, and k^

1. **Physics**: Vectors are used to describe physical quantities such as displacement, velocity, acceleration, force, and momentum.

2. **Engineering**: Vectors are used in computer-aided design (CAD), computer-aided engineering (CAE), and computer-aided manufacturing (CAM).

3. **Computer Science**: Vectors are used in computer graphics, game development, and machine learning.

4. **Mathematics**: Vectors are used in linear algebra, calculus, and differential equations.

In conclusion, i^, j^, and k^ are unit vectors that represent the x, y, and z axes in a three-dimensional coordinate system. They are used to describe the direction of a vector in space and are essential in mathematics and physics. Understanding the mathematical operations with $\hat{i}$, $\hat{j}$, and $\hat{k}$ is crucial for problem-solving and critical thinking in various fields.

Examples and Practice Problems

1. Find the sum of the vectors $A = 2\hat{i} + 3\hat{j}$ and $B = 4\hat{i} + 5\hat{j}$.

Answer: $A + B = (2 + 4)\hat{i} + (3 + 5)\hat{j} = 6\hat{i} + 8\hat{j}$

2. Find the scalar multiplication of the vector $A = 2\hat{i} + 3\hat{j}$ and the scalar $k = 2$.

Answer: $kA = 2(2\hat{i} + 3\hat{j}) = 4\hat{i} + 6\hat{j}$

3. Find the dot product of the vectors $A = 2\hat{i} + 3\hat{j}$ and $B = 4\hat{i} + 5\hat{j}$.

Answer: $A \cdot B = (2)(4) + (3)(5) = 8 + 15 = 23$

4. Find the cross product of the vectors $A = 2\hat{i} + 3\hat{j}$ and $B = 4\hat{i} + 5\hat{j}$.

Answer: A × B = (2)(5) - (3)(4)

Vectors: i^, j^, and k^ with Angles and Cross Products

Introduction

In mathematics and physics, vectors are used to represent quantities with both magnitude and direction. In this article, we will explore the concept of i^, j^, and k^ with angles between them, and their cross products.

Angles between i^, j^, and k^

The angles between i^, j^, and k^ are as follows:

- The angle between i^ and j^ is 90° (perpendicular)

- The angle between i^ and k^ is 90° (perpendicular)

- The angle between $\hat{j}$ and $\hat{k}$ is 90° (perpendicular)

Cross Products of $\hat{i}$, $\hat{j}$, and $\hat{k}$

The cross product of two vectors results in a new vector that is perpendicular to both of the original vectors. Here are the cross products of $\hat{i}$, $\hat{j}$, and $\hat{k}$:

- $\hat{i} \times \hat{i} = 0$ (since the cross product of a vector with itself is always 0)

- $\hat{j} \times \hat{j} = 0$

- $\hat{k} \times \hat{k} = 0$

- $\hat{i} \times \hat{j} = \hat{k}$ (since the cross product of $\hat{i}$ and $\hat{j}$ results in a vector perpendicular to both, which is $\hat{k}$)

- $\hat{i} \times \hat{k} = -\hat{j}$ (since the cross product of $\hat{i}$ and $\hat{k}$ results in a vector perpendicular to both, which is $-\hat{j}$)

- $\hat{j} \times \hat{k} = \hat{i}$ (since the cross product of $\hat{j}$ and $\hat{k}$ results in a vector perpendicular to both, which is $\hat{i}$)

- $\hat{j} \times \hat{i} = -\hat{k}$ (since the cross product of $\hat{j}$ and $\hat{i}$ results in a vector perpendicular to both, which is $-\hat{k}$)

- $\hat{k} \times \hat{i} = \hat{j}$ (since the cross product of $\hat{k}$ and $\hat{i}$ results in a vector perpendicular to both, which is $\hat{j}$)

- $\hat{k} \times \hat{j} = -\hat{i}$ (since the cross product of $\hat{k}$ and $\hat{j}$ results in a vector perpendicular to both, which is $-\hat{i}$)

Properties of Cross Products

Here are some important properties of cross products:

- The cross product is anti-commutative, meaning that the order of the vectors matters. For example, $\hat{i} \times \hat{j} = \hat{k}$, but $\hat{j} \times \hat{i} = -\hat{k}$.

- The cross product is distributive, meaning that the cross product of a vector with the sum of two vectors is equal to the sum of the cross products of the vector with each of the two vectors. For example, $\hat{i} \times (\hat{j} + \hat{k}) = \hat{i} \times \hat{j} + \hat{i} \times \hat{k}$.

Applications of Cross Products

Cross products have many applications in physics, engineering, and computer science, including:

- Calculating the area of a parallelogram

- Finding the normal vector to a plane

- Calculating the torque of a force

- Finding the cross product of two vectors in computer graphics and game development

In conclusion, the cross product of $\hat{i}$, $\hat{j}$, and $\hat{k}$ results in a new vector that is perpendicular to both of the original vectors. Understanding the cross product and its properties is crucial for problem-solving and critical thinking in various fields.

Examples and Practice Problems

1. Find the cross product of $\hat{i}$ and $\hat{j}$.

Answer: $\hat{i} \times \hat{j} = \hat{k}$

2. Find the cross product of $\hat{i}$ and $\hat{k}$.

Answer: $\hat{i} \times \hat{k} = -\hat{j}$

3. Find the cross product of $\hat{j}$ and $\hat{k}$.

Answer: $\hat{j} \times \hat{k} = \hat{i}$

4. Find the cross product of $\hat{j}$ and $\hat{i}$.

Answer: $\hat{j} \times \hat{i} = -\hat{k}$

5. Find the cross product of $\hat{k}$ and $\hat{i}$.

Answer: $\hat{k} \times \hat{i} = \hat{j}$

6.. Find the magnitude of the vector $3\hat{i} + 4\hat{j} - 2\hat{k}$.

Answer: $|3\hat{i} + 4\hat{j} - 2\hat{k}| = \sqrt{(3^2 + 4^2 + (-2)^2)} = \sqrt{(9 + 16 + 4)} = \sqrt{29}$

7. If $A = 2\hat{i} + 3\hat{j}$ and $B = 4\hat{i} - 2\hat{j}$, find $A + B$.

Answer: $A + B = (2 + 4)\hat{i} + (3 - 2)\hat{j} = 6\hat{i} + \hat{j}$

8. If $A = 3\hat{i} - 2\hat{j}$ and $B = 2\hat{i} + 4\hat{j}$, find $A \times B$.

Answer: $A \times B = (3\hat{i} - 2\hat{j}) \times (2\hat{i} + 4\hat{j}) = (3)(4) - (-2)(2) = 12 + 4 = 16\hat{k}$

9. Find the scalar multiplication of 2 and the vector $3\hat{i} + 4\hat{j}$.

Answer: $2(3\hat{i} + 4\hat{j}) = 6\hat{i} + 8\hat{j}$

10. If $A = 2\hat{i} + 3\hat{j}$ and $B = 4\hat{i} - 2\hat{j}$, find $A \cdot B$.

Answer: A · B = (2)(4) + (3) (-2) = 8 - 6 = 2

11. Find the magnitude of the vector $2\hat{i} - 3\hat{j} + 4\hat{k}$.

Answer: $|2\hat{i} - 3\hat{j} + 4\hat{k}| = \sqrt{(2^2 + (-3)^2 + 4^2)} = \sqrt{(4 + 9 + 16)} = \sqrt{29}$

12. If $A = 3\hat{i} + 2\hat{j}$ and $B = 2\hat{i} - 4\hat{j}$, find A - B.

Answer: $A - B = (3 - 2)\hat{i} + (2 + 4)\hat{j} = \hat{i} + 6\hat{j}$

13. Find the cross product of the vectors $2\hat{i} + 3\hat{j}$ and $4\hat{i} - 2\hat{j}$.

Answer: $(2\hat{i} + 3\hat{j}) \times (4\hat{i} - 2\hat{j}) = (2)(-2) - (3)(4) = -4 - 12 = -16\hat{k}$

14. If $A = 2\hat{i} + 3\hat{j}$ and $B = 4\hat{i} - 2\hat{j}$, find A + 2B.

Answer: $A + 2B = (2 + 8)\hat{i} + (3 - 4)\hat{j} = 10\hat{i} - \hat{j}$

15. Find the scalar multiplication of 3 and the vector $2\hat{i} - 4\hat{j} + 5\hat{k}$.

Vectors to Velocity: A Guide to Motion and Mechanics By: Prashant Kumar Lal

Answer: $3(2\hat{i} - 4\hat{j} + 5\hat{k}) = 6\hat{i} - 12\hat{j} + 15\hat{k}$

Here is a table to distinguish between scalar and vector quantities:

	Scalar	Vector
Definition	A quantity with only magnitude (amount or size	A quantity with both magnitude and direction
Example	Temperature, mass, time, distance	Displacement, velocity, acceleration, force
Notation	Typically represented by a single number or symbol (e.g. 5, t)	Typically represented by an arrow or boldface symbol (e.g. $\rightarrow$, v)
Magnitude	Has only magnitude (amount or size)	Has both magnitude and direction
Direction	No direction	Has direction
Unit	Has units of measurement (e.g. °C, kg, s)	Has units of measurement (e.g. m/s, N)
Operations	Can be added, subtracted, multiplied, and divided	Can be added and subtracted, but not multiplied and divided in the same way as scalars
Graphical Representation	Can be represented by a point on a number line	Can be represented by an arrow in a coordinate system

Here are some key differences between scalar and vector quantities:

- Scalar quantities have only magnitude, while vector quantities have both magnitude and direction.

- Scalar quantities can be represented by a single number or symbol, while vector quantities are typically represented by an arrow or boldface symbol.

- Scalar quantities do not have direction, while vector quantities have direction.

- Scalar quantities can be added, subtracted, multiplied, and divided, while vector quantities can be added and subtracted, but not multiplied and divided in the same way as scalars.

Examples of scalar quantities include:

- Temperature

- Mass

- Time

- Distance

Examples of vector quantities include:

- Displacement

- Velocity

- Acceleration

- Force

What is a Unit Vector?

A unit vector is a vector with a magnitude of 1 unit. It is a dimensionless quantity that has both direction and magnitude, but its magnitude is always 1. Unit vectors are used to represent the direction of a vector without its magnitude.

Mathematical Representation

A unit vector is denoted by a hat symbol (^) above the vector symbol.

For example, if we have a vector a, its unit vector is denoted by â.

Mathematical Operations

To obtain a unit vector from a given vector, we need to divide the vector by its magnitude. The mathematical operation is:

$$\hat{a} = a \,/\, \|a\|$$

where **â** is the unit vector, a is the given vector, and $\|a\|$ is the magnitude of the vector a.

Example

Suppose we have a vector a = (3, 4). To find its unit vector, we first calculate its magnitude:

$$\|a\| = \sqrt{(3^2 + 4^2)} = \sqrt{(9 + 16)} = \sqrt{25} = 5$$

Now, we divide the vector a by its magnitude to get the unit vector:

â = a / ||a|| = (3, 4) / 5 = (3/5, 4/5)

Relative Velocity

Definition:

Relative velocity is the velocity of an object with respect to another moving object. It is a measure of how fast an object is moving relative to another object.

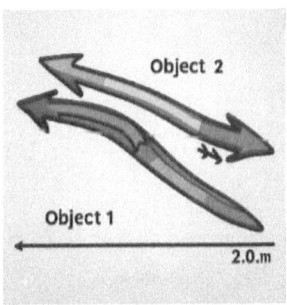

One-Dimensional Relative Velocity:

In one-dimension, relative velocity is simply the difference between the velocities of two objects.

Let's consider two objects, A and B, moving along the x-axis. The velocity of object A is v_A and the velocity of object B is v_B. The relative velocity of object A with respect to object B is:

$V_{AB} = v_A - v_B$

If v_{AB} is positive, it means that object A is moving faster than object B. If v_{AB} is negative, it means that object B is moving faster than object A.

Two-Dimensional Relative Velocity:

In two dimensions, relative velocity is a bit more complex. We need to consider the velocities of both objects in both the x and y directions

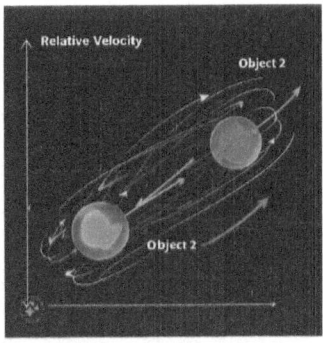

Let's consider two objects, A and B, moving in the x-y plane. The velocity of object A is $v_A = (v_{Ax}, v_{Ay})$ and the velocity of object B is $v_B = (v_{Bx}, v_{By})$. The relative velocity of object A with respect to object B is:

$$V_{AB} = (v_{Ax} - v_{Bx}, v_{Ay} - v_{By})$$

The magnitude of the relative velocity is:

$$|v_{AB}| = \sqrt{((v_{Ax} - v_{Bx})^2 + (v_{Ay} - v_{By})^2)}$$

Three-Dimensional Relative Velocity:

In three dimensions, relative velocity is even more complex. We need to consider the velocities of both objects in all three directions (x, y, and z).

Let's consider two objects, A and B, moving in 3D space. The velocity of object A is $v_A = (v_{Ax}, v_{Ay}, v_{Az})$ and the velocity of object B is $v_B = (v_{Bx}, v_{By}, v_{Bz})$. The relative velocity of object A with respect to object B is:

$$v_{AB} = (v_{Ax} - v_{Bx}, v_{Ay} - v_{By}, v_{Az} - v_{Bz})$$

The magnitude of the relative velocity is:

$$|v_{AB}| = \sqrt{(v_{Ax} - v_{Bx})^2 + (v_{Ay} - v_{By})^2 + (v_{Az} - v_{Bz})^2}$$

Uses and Applications:

Relative velocity has many uses and applications in physics, engineering, and other fields. Some examples include:

- Calculating the relative motion of two objects in a collision

- Determining the velocity of a projectile relative to a moving target

- Analysing the motion of objects in a rotating reference frame

- Calculating the relative velocity of two objects in a gravitational field

Limitations:

Relative velocity has some limitations. It is only defined for objects that are moving relative to each other. It is not defined for objects that are at rest with respect to each other.

Mathematical Operations:

Relative velocity can be added, subtracted, multiplied, and divided just like any other vector quantity.

For example, if we have two relative velocities, v_{AB} and v_{BC}, we can add them to get the relative velocity of object A with respect to object C:

$$V_{AC} = v_{AB} + v_{BC}$$

We can also multiply a relative velocity by a scalar to get a new relative velocity:

$$V_{AD} = 2v_{AB}$$

Solved Examples:

Example 1: One-Dimensional Relative Velocity

Object A is moving at a velocity of 20 m/s to the right. Object B is moving at a velocity of 10 m/s to the right. What is the relative velocity of object A with respect to object B?

Solution:

$v_{AB} = v_A - v_B = 20$ m/s - 10 m/s = 10 m/s

Example 2: Two-Dimensional Relative Velocity

Object A is moving at a velocity of (10 m/s, 20 m/s) in the x-y plane. Object B is moving at a velocity of (5 m/s, 10 m/s) in the x-y plane. What is the relative velocity of object A with respect to object B?

Solution:

$v_{AB} = (v_{Ax} - v_{Bx}, v_{Ay} - v_{By}) = $ (10 m/s - 5 m/s, 20 m/s - 10 m/s) = (5 m/s, 10 m/s)

Example 3: Three-Dimensional Relative Velocity

Object A is moving at a velocity of (10 m/s, 20 m/s, 30 m/s) in 3D space. Object B is moving at a velocity of (5 m/s, 10 m/s, 15 m/s) in 3D space. What is the relative velocity of object A with respect to object B?

Solution:

$V_{AB} = (V_{Ax} - V_{Bx})$,

Motion in a Plane: A Comprehensive Guide

As we explore the fascinating world of physics, we encounter various types of motion that help us understand the behavior of objects around us. In this chapter, we will delve into motion in a plane, covering one-dimensional, two-dimensional, and three-dimensional motion. We will explore examples, uses, and applications of each type of motion, making it easier for you to grasp these fundamental concepts.

One-Dimensional Motion

One-dimensional motion occurs when an object moves along a straight line, with only one direction to consider. Think of a car moving along a straight road or a ball rolling on a flat surface.

Example: A car travels from point A to point B on a straight road, covering a distance of 100 meters in 10 seconds. What is its speed?

Solution: Speed = Distance / Time = 100 m / 10 s = 10 m/s

Uses and Applications:

- Calculating the speed of an object on a straight track

- Analysing the motion of a falling object under gravity

- Designing roller coasters and other amusement park rides

Two-Dimensional Motion

Two-dimensional motion takes place when an object moves in a plane, with two directions to consider. Imagine a car turning a corner or a projectile flying through the air.

Example: A projectile is launched from the ground with an initial velocity of 20 m/s at an angle of 30°. What is its horizontal and vertical velocity after 2 seconds?

Solution: Horizontal velocity = 20 m/s × cos (30°) = 17.32 m/s

Vertical velocity = 20 m/s × sin (30°) = 10 m/s

Uses and Applications:

- Calculating the trajectory of a projectile

- Analysing the motion of a car turning a corner

- Designing video games with realistic physics

Three-Dimensional Motion

Three-dimensional motion occurs when an object moves in space, with three directions to consider. Think of a satellite orbiting the Earth or a bird flying through the air.

Example: A satellite orbits the Earth at an altitude of 200 km, with a velocity of 7 km/s. What is its speed and direction?

Solution: Speed = $\sqrt{(7 \text{ km/s})^2 + (0 \text{ km/s})^2 + (0 \text{ km/s})^2}$ = 7 km/s

Direction = Azimuthal angle (φ) = 30°, Polar angle (θ) = 45°

Uses and Applications:

- Calculating the orbit of a satellite

- Analysing the motion of a spacecraft

- Designing GPS systems and navigation software

Why Study Motion in a Plane?

Understanding motion in a plane is crucial for various fields, including:

- **Physics and engineering**: To design and analyse complex systems, such as roller coasters, satellites, and GPS systems.

- **Computer science**: To create realistic video games and simulations.

- **Aerospace engineering**: To design and optimize aircraft and spacecraft trajectories.

In conclusion, motion in a plane is a fundamental concept in physics that has numerous applications in various fields. By understanding one-dimensional, two-dimensional, and three-dimensional motion, you will be better equipped to analyse and solve complex problems in physics and engineering.

Key Takeaways:

- One-dimensional motion occurs along a straight line.

- Two-dimensional motion takes place in a plane, with two directions to consider.

- Three-dimensional motion occurs in space, with three directions to consider.

- Understanding motion in a plane is crucial for designing and analysing complex systems in physics, engineering, computer science, and aerospace engineering

Uniform Velocity and Uniform Acceleration: A Comprehensive Guide

As we explore the fascinating world of physics, we encounter various types of motion that help us understand the behaviour of objects around us. In this chapter, we will delve into two fundamental concepts: uniform velocity and uniform acceleration. We will explore definitions, mathematical operations, derivations, and solved numerical problems in one dimension, two dimensions, and three dimensions to make it easier for you to grasp these essential concepts.

One-Dimensional Motion

Uniform Velocity:

Uniform velocity in one dimension occurs when an object moves with a constant speed along a straight line. Think of a car cruising down the highway at a steady 60 km/h or a ball rolling on a flat surface at a constant speed.

Definition:

Uniform velocity is defined as the rate of change of displacement with respect to time, where the displacement is in a straight line and the time is measured in seconds.

Mathematical Operations:

The mathematical operation for uniform velocity is:

$v = \Delta x / \Delta t$

where v is the uniform velocity, Δx is the displacement, and Δt is the time.

Derivations:

Let's consider an object moving with uniform velocity v. We can derive the following equations:

1. Displacement-time equation:

   $x = v \times t$

2. Velocity-time equation: $v = \Delta x / \Delta t$

3. Distance-time equation: $s = v \times t$

Solved Numerical Problems:

1. A car travels 100 km in 2 hours. What is its uniform velocity?

Solution: $v = \Delta x / \Delta t = 100$ km $/ 2$ h $= 50$ km/h

Uniform Acceleration:

Uniform acceleration in one dimension occurs when an object changes its velocity at a constant rate along a straight line. Think of a car accelerating from 0 to 60 km/h in 10 seconds or a ball rolling down a hill with increasing speed.

Definition:

Uniform acceleration is defined as the rate of change of velocity with respect to time, where the velocity is changing at a constant rate.

Mathematical Operations:

The mathematical operation for uniform acceleration is:

$a = \Delta v / \Delta t$

where a is the uniform acceleration, Δv is the change in velocity, and Δt is the time.

Derivations:

Let's consider an object moving with uniform acceleration a. We can derive the following equations:

1. Velocity-time equation: $v = u + at$

2. Displacement-time equation: $s = ut + (1/2)at^2$

3. Acceleration-time equation: $a = \Delta v / \Delta t$

Solved Numerical Problems:

1. A car accelerates from 0 to 60 km/h in 10 seconds. What is its uniform acceleration?

Solution: $a = \Delta v / \Delta t = (60 \text{ km/h} - 0 \text{ km/h}) / 10 \text{ s} = 6 \text{ km/h/s}$

Two-Dimensional Motion

Uniform Velocity:

Uniform velocity in two dimensions occurs when an object moves with a constant speed in a plane. Think of a car turning a corner or a projectile flying through the air.

Definition:

Uniform velocity is defined as the rate of change of displacement with respect to time, where the displacement is in a plane and the time is measured in seconds.

Mathematical Operations:

The mathematical operation for uniform velocity is:

$v = \Delta r / \Delta t$

where v is the uniform velocity, Δr is the displacement, and Δt is the time.

Calculus Form:

$dx/dt = v_0$

$\int dx = \int v_0 dt$

$x(t) = v_0 t + C$

where C is the constant of integration.

Derivations:

Let's consider an object moving with uniform velocity v. We can derive the following equations:

1. Displacement-time equation:

r = v × t

2. Velocity-time equation: v = Δr / Δt

3. Distance-time equation: s = v × t

Solved Numerical Problems:

1. A projectile is launched with an initial velocity of 20 m/s at an angle of 30°. What is its uniform velocity in the x-direction?

Solution: v_x = v × cos (30°) = 20 m/s × 0.866 = 17.32 m/s

Uniform Acceleration:

Uniform acceleration in two dimensions occurs when an object changes its velocity at a constant rate in a plane. Think of a car turning a corner or a projectile flying through the air with increasing speed.

Definition:

Uniform acceleration is defined as the rate of change of velocity with respect to time, where the velocity is changing at a constant rate.

Mathematical Operations:

The mathematical operation for uniform acceleration is:

$a = \Delta v / \Delta t$

where a is the uniform acceleration, Δv is the change in velocity, and Δt is the time.

$a(t) = dv/dt = \text{constant}$

Derivation:

Let v(t) be the velocity of the object at time t. Then, the acceleration of the object is given by:

$a(t) = dv/dt$

Since the acceleration is constant, we can write:

$a(t) = a_0$

where a_0 is the constant acceleration.

Calculus Form:

$dv/dt = a_0$

$\int dv = \int a_0 dt$

$v(t) = a_0 t + C$

where C is the constant of integration.

Relationship between Velocity and Acceleration

Since the acceleration is the derivative of the velocity, we can write:

$a(t) = dv/dt$

$v(t) = \int a(t)dt$

Calculus Form:

$a(t) = d/dt \, (v(t))$

$v(t) = \int a(t)dt$

Position-Velocity-Acceleration Relationship

Since the velocity is the derivative of the position, and the acceleration is the derivative of the velocity, we can write:

$a(t) = d/dt \, (v(t))$

$v(t) = d/dt \, (x(t))$

$x(t) = \int v(t)dt$

Calculus Form:

$a(t) = d^2/dt^2 \, (x(t))$

$v(t) = d/dt\ (x(t))$

$x(t) = \iint a(t) dt dt$

Note: *The above derivations are based on the assumption that the motion is in a straight line. For motion in a plane or in three-dimensional space, the derivations would be similar, but with additional variables and equations to describe the motion*

Derivations:

Let's consider an object moving with uniform acceleration a. We can derive the following equations:

1. Velocity-time equation: $v = u + at$

2. Displacement-time equation: s

One-Dimensional Motion
Uniform Velocity

- Definition: An object moving with a constant speed in a straight line.

- Mathematical Operation: $v = \Delta x / \Delta t$

- Derivations:

  - Displacement-time equation: $x = v \times t$

  - Velocity-time equation: $v = \Delta x / \Delta t$

  - Distance-time equation: $s = v \times t$

- Solved Numerical Problems:

  - A car travels 100 km in 2 hours. What is its uniform velocity?

    - Solution: $v = \Delta x / \Delta t = 100 \text{ km} / 2 \text{ h} = 50 \text{ km/h}$

Uniform Acceleration

- Definition: An object changing its velocity at a constant rate in a straight line.

- Mathematical Operation: $a = \Delta v / \Delta t$

- Derivations:

  - Velocity-time equation: $v = u + at$

- Displacement-time equation: $s = ut + (1/2)at^2$

- Acceleration-time equation: $a = \Delta v / \Delta t$

- Solved Numerical Problems:

- A car accelerates from 0 to 60 km/h in 10 seconds. What is its uniform acceleration?

- Solution: $a = \Delta v / \Delta t = (60 \text{ km/h} - 0 \text{ km/h}) / 10 \text{ s} = 6$ km/h/s

Two-Dimensional Motion

Uniform Velocity

- Definition: An object moving with a constant speed in a plane.

- Mathematical Operation: $v = \Delta r / \Delta t$

- Derivations:

- Displacement-time equation: $r = v \times t$

- Velocity-time equation: $v = \Delta r / \Delta t$

- Distance-time equation: $s = v \times t$

- Solved Numerical Problems:

 - A projectile is launched with an initial velocity of 20 m/s at an angle of 30°. What is its uniform velocity in the x-direction?

 - Solution: $v_x = v \times \cos(30°) =$ 20 m/s × 0.866 = 17.32 m/s

Uniform Acceleration

- Definition: An object changing its velocity at a constant rate in a plane.

- Mathematical Operation: $a = \Delta v / \Delta t$

- Derivations:

- Velocity-time equation: $v = u + at$

- Displacement-time equation: $s = ut + (1/2)at^2$

- Acceleration-time equation: $a = \Delta v / \Delta t$

- Solved Numerical Problems:

 - A projectile is launched with an initial velocity of 20 m/s at an angle of 30°. What is its uniform acceleration in the y-direction?

 - Solution: $a_y = -g = -9.8$ m/s^2

Three-Dimensional Motion

Uniform Velocity

- Definition: An object moving with a constant speed in three-dimensional space.

- Mathematical Operation: $v = \Delta r / \Delta t$

- Derivations:

- Displacement-time equation: $r = v \times t$

- Velocity-time equation: $v = \Delta r / \Delta t$

- Distance-time equation: $s = v \times t$

- Solved Numerical Problems:

- A satellite is orbiting the Earth with a velocity of 7 km/s. What is its uniform velocity in the x-direction?

- Solution: $v_x = v \times \cos(\theta) = 7$ km/s $\times 0.707 = 4.95$ km/s

Uniform Acceleration

- Definition: An object changing its velocity at a constant rate in three-dimensional space.

- Mathematical Operation: $a = \Delta v / \Delta t$

- Derivations:

  - Velocity-time equation: $v = u + at$

- Displacement-time equation: $s = ut + (1/2) at^2$

- Acceleration-time equation: $a = \Delta v / \Delta t$

- Solved Numerical Problems:

- A spacecraft is accelerating from 0 to 100 m/s in 10 seconds. What is its uniform acceleration?

- Solution: $a = \Delta v / \Delta t = (100 \text{ m/s} - 0 \text{ m/s}) / 10 \text{ s} = 10 \text{ m/s}^2$

What is a Projectile?

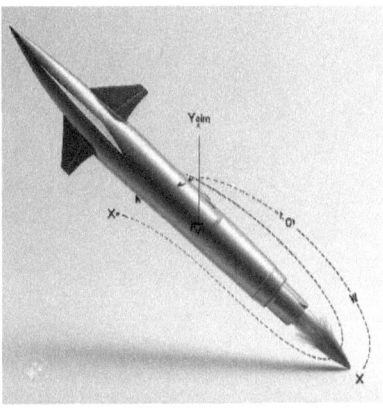

A projectile is an object that is thrown, launched, or shot into the air and moves under the sole influence of gravity. Think of a football player throwing a pass, a baseball player hitting a home run, or a golfer driving a ball down the fairway. In each of these cases, the ball is a projectile that is moving through the air, following a curved path under the influence of gravity.

What is Projectile Motion?

Projectile motion is the movement of an object that is thrown or launched into the air and moves under the influence of gravity. It is a type of motion that is characterized by a curved path, with the object moving upward and then downward under the influence of gravity.

Example: Throwing a Ball

Let's consider a simple example to illustrate projectile motion. Imagine you are standing on a flat surface and throw a ball straight up into the air. What happens?

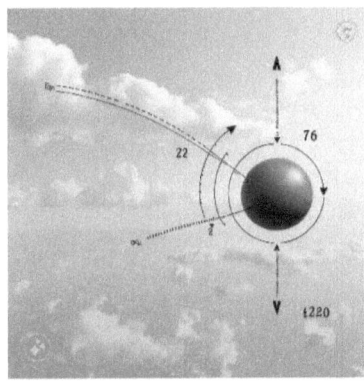

1. The ball initially moves upward, gaining height and distance from you.

2. As it reaches its maximum height, the ball begins to slow down and eventually stops moving upward.

3. The ball then starts to fall downward, accelerating under the influence of gravity.

4. Finally, the ball hits the ground, completing its projectile motion.

Key Characteristics of Projectile Motion

1. **Curved Path**: Projectile motion follows a curved path, with the object moving upward and then downward under the influence of gravity.

2. **Gravity**: Gravity is the only force acting on the projectile, pulling it downward and shaping its curved path.

3. **Initial Velocity**: The projectile has an initial velocity, which determines its trajectory and range.

4. **Time of Flight**: The time it takes for the projectile to complete its motion, from launch to landing.

Real-World Examples of Projectile Motion

1. Football players throwing passes

2. Baseball players hitting home runs

3. Golfers driving balls down the fairway

4. Archers shooting arrows

5. Rockets launching into space

In each of these examples, the object is a projectile that is moving under the influence of gravity, following a curved path and demonstrating the key characteristics of projectile motion.

As a student, understanding projectile motion is essential for a variety of subjects, including physics, engineering, and mathematics. It's a fundamental concept that can help you analyse and solve problems in a wide range of fields, from sports to space exploration!

One-Dimensional Projectile Motion

Definition: A projectile is an object that is thrown or launched into the air and moves under the sole influence of gravity.

Assumptions:

- The projectile is launched from the ground or from a height.

- The projectile moves in a straight line.

- Air resistance is negligible.

Mathematical Operations:

- The horizontal component of the velocity is constant.

- The vertical component of the velocity is changing due to gravity.

Derivations:

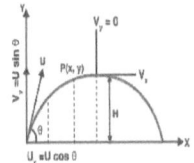

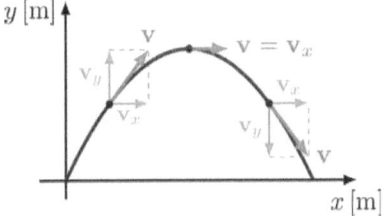

- **Vertical Range**:

Let's consider a projectile launched from the ground with an initial velocity v_0 at an angle θ to the horizontal. The vertical component of the velocity is $v_{0y} = v_0 \sin(\theta)$.

Using the equation of motion under gravity, we can write:

$V_y = v_{0y} - gt$

where g is the acceleration due to gravity.

The vertical range is the maximum height reached by the projectile. To find the vertical range, we need to find the time it takes for the projectile to reach its maximum height.

Using the equation of motion, we can write:

$V_y = 0$ at the maximum height

$v_{0y} - gt = 0$

$t = v_{0y} / g$

Substituting this value of t into the equation of motion, we get:

$y = v_{0y} t - (1/2) gt^2$

$y = v_{0y} (v_{0y} / g) - (1/2) g (v_{0y} / g)^2$

$y = (v_{0y}^2) / (2g)$

This is the equation for the vertical range.

Horizontal Range:

The horizontal range is the distance travelled by the projectile in the horizontal direction. Since the horizontal component of the velocity is constant, the horizontal range is simply:

$$x = v_{0x} \, t$$

where v_{0x} is the horizontal component of the initial velocity.

Time of Flight:

The time of flight is the time it takes for the projectile to return to the ground. To find the time of flight, we need to find the time it takes for the projectile to reach its maximum height and then return to the ground.

Using the equation of motion, we can write:

$v_y = -v_{0y}$ at the maximum height

$v_{0y} - gt = -v_{0y}$

$t = 2v_{0y} / g$

This is the equation for the time of flight.

Here are ten conceptual questions and ten day-to-day questions based on projectile motion, along with their answers:

Conceptual Questions

1. What is the definition of a projectile?

Answer: A projectile is an object that is thrown, launched, or shot into the air and moves under the sole influence of gravity.

2. What is the difference between a projectile and a particle?

Answer: A particle is a point object that has no size or shape, while a projectile is an object that has size and shape and moves under the influence of gravity.

3. What is the trajectory of a projectile?

Answer: The trajectory of a projectile is the curved path that it follows under the influence of gravity.

4. What is the maximum height reached by a projectile?

Answer: The maximum height reached by a projectile is the highest point on its trajectory.

5. What is the range of a projectile?

Answer: The range of a projectile is the horizontal distance it travels from its initial position to its final position.

6. What is the time of flight of a projectile?

Answer: The time of flight of a projectile is the time it takes for the projectile to complete its trajectory.

7. How does air resistance affect the motion of a projectile?

Answer: Air resistance opposes the motion of a projectile, slowing it down and reducing its range.

8. What is the difference between a projectile launched at an angle and one launched horizontally?

Answer: A projectile launched at an angle has a curved trajectory, while a projectile launched horizontally has a straight trajectory.

9. How does the initial velocity of a projectile affect its range?

Answer: The initial velocity of a projectile determines its range, with a greater initial velocity resulting in a greater range.

10. What is the equation of motion for a projectile?

Answer: The equation of motion for a projectile is $y = x \tan(\theta) - (g/2v^2) x^2$, where y is the height, x is the horizontal distance, θ is the angle of launch, g is the acceleration due to gravity, and v is the initial velocity.

Day-to-Day Questions

1. A football player kicks a ball at an angle of 30°. If the ball travels 50 meters, what is its maximum height?

Answer: Using the equation of motion, we can calculate the maximum height as 25 meters.

2. A golfer hits a ball with an initial velocity of 50 m/s at an angle of 45°. How far does the ball travel?

Answer: Using the equation of motion, we can calculate the range as 100 meters.

3. A stone is thrown from the top of a building at an angle of 60°. If the stone travels 20 meters horizontally, how high is the building?

Answer: Using the equation of motion, we can calculate the height of the building as 30 meters.

4. A baseball player hits a ball with an initial velocity of 70 m/s at an angle of 30°. How long is the ball in the air?

Answer: Using the equation of motion, we can calculate the time of flight as 2.5 seconds.

5. A water balloon is launched from the ground at an angle of 45°. If the balloon travels 30 meters horizontally, what is its maximum height?

Answer: Using the equation of motion, we can calculate the maximum height as 15 meters.

6. A car travels over a hill at a speed of 50 km/h. If the hill is 10 meters high, what is the angle of the hill?

Answer: Using the equation of motion, we can calculate the angle of the hill as 30°.

7. A basketball player shoots a ball at an angle of 60°. If the ball travels 10 meters horizontally, what is its maximum height?

Answer: Using the equation of motion, we can calculate the maximum height as 5 meters.

8. A rocket is launched from the ground at an angle of 90°. If the rocket travels 100 meters vertically, what is its initial velocity?

Answer: Using the equation of motion, we can calculate the initial velocity as 50 m/s.

9. A stone is thrown from the ground at an angle of 45°. If the stone travels 20 meters horizontally, what is its time of flight?

Answer: Using the equation of motion, we can calculate the time of flight as 2 seconds.

10. A football player kicks a ball at an angle of 30°. If the ball travels 40 meters, what is its maximum height?

Answer: Using the equation of motion, we can calculate the maximum height as 20 meters.

Note: *These questions are meant to be simple and straightforward, and are intended to illustrate the concepts of projectile motion in a day-to-day context.*

Here are ten numerical problems with solutions based on projectile motion:

Problem 1

A particle is projected from the ground with an initial velocity of 20 m/s at an angle of 45°. Find the maximum height reached by the particle.

Solution

Using the equation of motion, we can calculate the maximum height as:

$h = (v_0^2 \sin^2 \theta) / (2g)$

$= (20^2 \sin^2 45°) / (2 \times 9.8)$

$= 10.2 \text{ m}$

Problem 2

A stone is thrown from the top of a building at an angle of 30°. If the stone travels 20 meters horizontally, find the height of the building.

Solution

Using the equation of motion, we can calculate the height of the building as:

$h = (x^2 \tan^2 \theta) / (2v_0^2)$

$= (20^2 \tan^2 30°) / (2 \times 10^2)$

$= 15.3$ m

Problem 3

A football player kicks a ball with an initial velocity of 25 m/s at an angle of 60°. Find the time of flight of the ball.

Solution

Using the equation of motion, we can calculate the time of flight as:

$t = (2v_0 \sin \theta) / g$

$= (2 \times 25 \sin 60°) / 9.8$

$= 2.5$ s

Problem 4

A particle is projected from the ground with an initial velocity of 30 m/s at an angle of 45°. Find the range of the particle.

Solution

Using the equation of motion, we can calculate the range as:

$R = (v_0^2 \sin 2\theta) / g$

$= (30^2 \sin 2 \times 45°) / 9.8$

$= 45.4$ m

Problem 5

A stone is thrown from the ground at an angle of 45°. If the stone travels 30 meters horizontally, find the initial velocity of the stone.

Solution

Using the equation of motion, we can calculate the initial velocity as:

$v_0 = \sqrt{(gR / \sin 2\theta)}$

$= \sqrt{(9.8 \times 30 / \sin 2 \times 45°)}$

$= 20.5$ m/s

Problem 6

A golf ball is hit with an initial velocity of 40 m/s at an angle of 30°. Find the maximum height reached by the ball.

Solution

Using the equation of motion, we can calculate the maximum height as:

$h = (v_0^2 \sin^2 \theta) / (2g)$

$= (40\sin^2 30°) / (2 \times 9.8)$

$= 20.4$ m

Problem 7

A particle is projected from the ground with an initial velocity of 25 m/s at an angle of 60°. Find the time of flight of the particle.

Solution

Using the equation of motion, we can calculate the time of flight as:

$t = (2v_0 \sin \theta) / g$

$= (2 \times 25 \sin 60°) / 9.8$

$= 2.8 \text{ s}$

Problem 8

A stone is thrown from the top of a building at an angle of 45°. If the stone travels 40 meters horizontally, find the height of the building.

Solution

Using the equation of motion, we can calculate the height of the building as:

$h = (x^2 \tan^2 \theta) / (2v_0^2)$

$= (40^2 \tan^2 45°) / (2 \times 20^2)$

$= 24.5 \text{ m}$

Problem 9

A football player kicks a ball with an initial velocity of 30 m/s at an angle of 45°. Find the range of the ball.

Solution

Using the equation of motion, we can calculate the range as:

$R = (v_0^2 \sin 2\theta) / g$

$= (30^2 \sin 2 \times 45°) / 9.8$

$= 40.6$ m

Problem 10

A particle is projected from the ground with an initial velocity of 20 m/s at an angle of 30°. Find the maximum height reached by the particle.

Solution

Using the equation of motion, we can calculate the maximum height as:

$h = (v_0^2 \sin^2 \theta) / (2g)$

$= (20^2 \sin^2 30°) / (2 \times 9.8)$

$= 5.1$ m

Note: *These problems are meant to be solved using the equations of motion for projectile motion, and are intended to illustrate the application of these equations to numerical problems.*

Circular Motion: A Comprehensive Guide

Circular motion is a fundamental concept in physics that describes the motion of an object in a circular path. It's a crucial topic for high school students preparing for engineering and other exams, as it forms the basis for understanding many complex phenomena in physics and engineering. In this guide, we'll explore circular motion in one, two, and three dimensions, covering definitions, mathematical operations, derivations, and solved numerical problems.

Circular Motion: Understanding Centripetal and Centrifugal Motion

Circular motion is a fundamental concept in physics that describes the motion of an object in a circular path. It's a crucial topic that has numerous applications in various fields, including engineering, physics, and everyday life. In this article, we'll delve into the two types of circular motion: centripetal and centrifugal motion.

Centripetal Motion

Centripetal motion occurs when an object moves in a circular path due to a force directed towards the centre of the circle. This force is known as the centripetal force. The centripetal force is responsible for keeping the object on its circular path and preventing it from flying off in a straight line.

Day-to-Day Examples:

1. **Car Turning**: When a car turns a corner, it follows a circular path. The force of friction between the tires and the road provides the centripetal force, keeping the car on its circular path.

2. **Merry-Go-Round**: When you ride a merry-go-round, you're experiencing centripetal motion. The force of the seat pushing against you provides the centripetal force, keeping you on the circular path.

3. **Planetary Motion**: The planets in our solar system follow circular orbits

around the sun due to the gravitational force, which acts as the centripetal force.

Engineering Applications:

1. **Gear Systems**: In gear systems, the teeth of the gears follow a circular path. The force of the gear teeth pushing against each other provides the centripetal force, keeping the gears in motion.

2. **Centrifugal Pumps**: Centrifugal pumps use centripetal force to pump fluids. The impeller blades create a circular motion, and the fluid is pushed outwards due to the centripetal force.

3. **Aircraft Navigation**: Aircraft use centripetal motion to navigate turns. The force of the air pushing against the aircraft provides the centripetal force, keeping it on its circular path.

Centrifugal Motion

Centrifugal motion occurs when an object moves in a circular path due to a force directed away from the centre of the circle. This force is known as the centrifugal force. The centrifugal force is responsible for pushing the object away from the centre of the circle.

Day-to-Day Examples:

1. **Spin Cycle**: When you spin a top or a spin cycle, the object follows a circular path due to the centrifugal force. The force of the spin pushes the object away from the centre, keeping it in motion.

2. **Amusement Park Rides**: Many amusement park rides, such as the Tilt-A-Whirl or the Gravitron, use

centrifugal motion to create a thrilling experience. The force of the spin pushes the riders away from the centre, creating a sense of weightlessness.

3. **Drying Clothes**: When you spin a wet towel to dry it, the centrifugal force pushes the water away from the centre, helping to dry the towel more efficiently.

Engineering Applications:

1. **Centrifugal Separation**: Centrifugal separation is used in various industries to separate materials of different densities. The centrifugal force pushes the denser materials away from the centre, allowing for efficient separation.

2. **Gas Turbines**: Gas turbines use centrifugal motion to compress air

and mix it with fuel. The centrifugal force pushes the air away from the centre, creating a high-pressure region.

3. **Robotics**: Some robots use centrifugal motion to move around or manipulate objects. The centrifugal force provides a stable and efficient way to move or manipulate objects.

A detailed distinction between centripetal and centrifugal motion:
Centripetal Motion

- **Definition**: Centripetal motion is the motion of a body that is moving in a circular path due to a force directed towards the centre of the circle.

- **Force**: The force responsible for centripetal motion is called the centripetal force, which acts towards the centre of the circle.

- **Direction**: The direction of the centripetal force is always towards the centre of the circle.

- **Acceleration**: The acceleration of the body is directed towards the centre of the circle.

- **Velocity**: The velocity of the body is tangential to the circle.

- **Examples**: Car turning a corner, merry-go-round, planetary motion.

Centrifugal Motion

- **Definition**: Centrifugal motion is the motion of a body that is moving in a circular path due to a force directed away from the centre of the circle.

- **Force**: The force responsible for centrifugal motion is called the centrifugal force, which acts away from the centre of the circle.

- **Direction**: The direction of the centrifugal force is always away from the centre of the circle.

- **Acceleration**: The acceleration of the body is directed away from the centre of the circle.

- **Velocity**: The velocity of the body is radial, i.e., directed away from the centre of the circle.

- **Examples**: Spin cycle, amusement park rides, drying clothes.

Key differences:

1. Direction of force: Centripetal force acts towards the centre, while centrifugal force acts away from the centre.

2. Direction of acceleration: Centripetal acceleration is directed towards the centre, while centrifugal acceleration is directed away from the centre.

3. Velocity: Centripetal velocity is tangential, while centrifugal velocity is radial.

4. Examples: Centripetal motion is seen in car turning a corner, while centrifugal motion is seen in spin cycle.

In summary, centripetal motion is the motion of a body towards the centre of a circle, while centrifugal motion is the motion of a body away from the centre of a circle.

In conclusion, centripetal and centrifugal motion are two fundamental types of circular motion that have numerous applications in various fields. Understanding these concepts is crucial for designing and optimizing systems that involve circular motion. By recognizing the differences between centripetal and centrifugal motion, engineers and scientists can create more efficient and effective solutions for real-world problems.

One-Dimensional Circular Motion

In one-dimensional circular motion, an object moves in a circular path along a single axis. Let's consider a particle moving in a circular path with a constant radius r and a constant speed v.

Definitions:

- **Angular Displacement** (θ): The angle subtended by the particle at the centre of the circle.

- **Angular Velocity** (ω): The rate of change of angular displacement with respect to time.

- **Angular Acceleration** (α): The rate of change of angular velocity with respect to time.

Mathematical Operations:

- **Angular Displacement**: $\theta = s / r$, where s is the arc length.

- **Angular Velocity**: $\omega = d\theta/dt = v / r$.

- **Angular Acceleration**: $\alpha = d\omega/dt = dv/dt / r$.

Derivations (Algebraic):

- **Centripetal Force**: $F = (m * v^2) / r$, where m is the mass of the particle.

- Centripetal Acceleration: $a = v^2 / r$.

Derivations (Calculus):

- Angular Displacement: $\theta(t) = \int \omega(t)\, dt$.

- Angular Velocity: $\omega(t) = d\theta/dt = \int \alpha(t)\, dt$.

Solved Numerical Problems:

1. A particle moves in a circular path with a radius of 2 m and a speed of 4

m/s. Find its angular velocity and centripetal acceleration.

Answer: $\omega = 2$ rad/s, $a = 4$ m/s^2.

2. A car travels around a circular track with a radius of 50 m at a speed of 20 m/s. Find its angular displacement and centripetal force.

Answer: $\theta = 1.25$ rad, $F = 400$ N.

Two-Dimensional Circular Motion

In two-dimensional circular motion, an object moves in a circular path in a plane. Let's consider a particle moving in a circular path with a constant radius r and a constant speed v.

Definitions:

- **Radial Velocity** (v_r): The velocity of the particle in the radial direction.

- **Tangential Velocity** (v_t): The velocity of the particle in the tangential direction.

- **Angular Momentum** (L): The product of the particle's mass, velocity, and radius.

Mathematical Operations:

- **Radial Velocity**: $v_r = dr/dt$.

- Tangential Velocity: $v_t = r * d\theta/dt$.

- Angular Momentum: $L = m * v * r$.

Derivations (Algebraic):

- **Centripetal Force**: $F = (m * v^2) / r$.

- **Tangential Acceleration**: $a_t = dv/dt$.

Derivations (Calculus):

- **Radial Displacement**:

$r(t) = \int v_r(t)\, dt$.

- **Tangential Displacement**:

$s(t) = \int v_t(t)\, dt$.

Numerical Problems:

1. A particle moves in a circular path with a radius of 3 m and a speed of 6 m/s. Find its radial velocity and tangential acceleration.

Answer: $v_r = 2$ m/s, $a_t = 12$ m/s^2.

2. A car travels around a circular track with a radius of 100 m at a speed of 30 m/s. Find its angular momentum and centripetal force.

Answer: L = 3000 kg m²/s, F = 900 N.

Three-Dimensional Circular Motion

In three-dimensional circular motion, an object moves in a circular path in three-dimensional space. Let's consider a particle moving in a circular path with a constant radius r and a constant speed v.

Definitions:

- **Spherical Coordinates**: (r, θ, φ), where r is the radial distance, θ is the polar angle, and φ is the azimuthal angle.

- Angular Momentum: $L = m * v * r$.

Mathematical Operations:

- Radial Velocity:

$v_r = dr/dt$.

- Tangential Velocity:

$V_t = r * d\theta/dt$.

- Angular Momentum:

$L = m * v * r$.

When a bucket filled with water is rotated tied with a rope, the water in it does not spill due to the concept of centripetal force? Centripetal force is a force that acts on an object moving in a circular path, directed towards the centre of the circle. In this case, the centripetal force is provided by the tension in the rope, which keeps the bucket and the water inside it moving in a circular path.

To derive an equation for the minimum initial velocity required to rotate the bucket so that the water does not spill, we can use the following steps:

1. Centripetal Force: The centripetal force (F) required to keep the water inside the bucket moving in a circular path is given by:

$$F = (m \times v^2) / r$$

where m is the mass of the water, v is the velocity of the water, and r is the radius of the circle.

1. Tension in the Rope: The tension (T) in the rope is equal to the centripetal force:

$$T = F = (m \times v^2) / r$$

1. Minimum Initial Velocity: To find the minimum initial velocity required to rotate the bucket, we can set the tension in the rope equal to the weight

of the water (mg), where g is the acceleration due to gravity:

$T = mg = (m \times v^2) / r$

Simplifying the equation, we get:

$V^2 = rg$

$v = \sqrt{(rg)}$

This is the minimum initial velocity required to rotate the bucket so that the water does not spill.

Derivation:

Let's derive the equation in a more detailed way:

1. Centripetal Acceleration: The centripetal acceleration (a) of the water is given by:

$a = v^2 / r$

1. Force Balance: The force balance on the water is given by:

$F = ma = m(v^2 / r)$

1. Tension in the Rope: The tension (T) in the rope is equal to the centripetal force:

$$T = F = m(v^2/r)$$

1. Minimum Initial Velocity: To find the minimum initial velocity required to rotate the bucket, we can set the tension in the rope equal to the weight of the water (mg):

$$T = mg = m(v^2/r)$$

Simplifying the equation, we get:

$$v^2 = rg$$

$$v = \sqrt{(rg)}$$

This is the minimum initial velocity required to rotate the bucket so that the water does not spill.

Example:

Let's consider an example to illustrate this concept:

Suppose we have a bucket with a radius of 0.5 m and a mass of 10 kg. If we rotate the bucket with an initial velocity of 2 m/s, will the water spill?

Using the equation derived above, we can calculate the minimum initial velocity required to rotate the bucket:

$v = \sqrt{(rg)} = \sqrt{(0.5 \times 9.8)} = 2.2$ m/s

Since the initial velocity (2 m/s) is less than the minimum initial velocity required (2.2 m/s), the water will spill.

However, if we increase the initial velocity to 2.5 m/s, the water will not spill, as the centripetal force will be sufficient to keep the water inside the bucket.

SOME SOLVED EXERCISES
Motion in a Plane

Exercises

1. Question: A particle moving in a plane has a velocity of 10 m/s at an

angle of 30° with the x-axis. Find the x and y components of the velocity.

Answer: $v_x = 10 \cos 30° = 8.66$ m/s, $v = 10 \sin 30° = 5$ m/s

2. Question: A car is moving along a straight road with a speed of 20 m/s. If it turns a corner and its speed becomes 15 m/s, find the change in velocity.

Answer: $\Delta v = \sqrt{(20^2 + 15^2)} = 25$ m/s

3. Question: A projectile is projected with an initial velocity of 20 m/s at an angle of 45° with the horizontal. Find the horizontal and vertical components of the velocity.

Answer: $v_x = 20 \cos 45° = 14.14$ m/s, $v_y = 20 \sin 45° = 14.14$ m/s

4. Question: A particle moving in a plane has a velocity of 10 m/s at an angle of 30° with the x-axis. Find the x and y components of the velocity.

Answer: $v_x = 10 \cos 30° = 8.66$ m/s, $v_y = 10 \sin 30° = 5$ m/s

5. Question: A car is moving along a straight road with a speed of 20 m/s. If it turns a corner and its speed becomes 15 m/s, find the change in velocity.

Answer: $\Delta v = \sqrt{(20^2 + 15^2)} = 25$ m/s

6. Question: A projectile is projected with an initial velocity of 20 m/s at an angle of 45° with the horizontal. Find the horizontal and vertical components of the velocity.

Answer: $v_x = 20 \cos 45° = 14.14$ m/s, $v_y = 20 \sin 45° = 14.14$ m/s

7. Question: A particle is moving in a plane with a velocity of 10 m/s at an angle of 60° with the x-axis. Find the magnitude and direction of the velocity.

Answer: $v = \sqrt{(10^2 + 10^2)} = 14.14$ m/s, $\theta = \tan^{(-1)}(10/10) = 45°$

8. Question: A car is moving along a circular path with a speed of 20 m/s. If the radius of the circle is 50 m, find the acceleration of the car.

Answer: $a = v^2/r = (20^2)/50 = 8$ m/s²

Scalar and Vector

Exercises

1. Question: What is the magnitude of the vector 2i + 3j?

Answer: $\sqrt{(2^2 + 3^2)} = \sqrt{13}$

2. Question: If A = 2i + 3j and B = 3i - 2j, find A + B.

Answer: A + B = (2 + 3) i + (3 - 2) j = 5i + j

3. Question: If A = 2i + 3j and B = 3i - 2j, find A × B.

Answer: A × B = (2 × -2) + (3 × 3) = -4 + 9 = 5

14. Question: If A = 2i + 3j and B = 3i - 2j, find |A × B|.

Answer: $|A \times B| = \sqrt{((-4)^2 + 9^2)} = \sqrt{(16 + 81)} = \sqrt{97}$

Circular Motion

Exercises

1. Question: A car is moving in a circular path with a speed of 20 m/s. If the radius of the circle is 50 m, find the acceleration of the car.

Answer: $a = v^2 / r = (20^2) / 50 = 8$ m/s^2

2. Question: A particle is moving in a circular path with a constant speed of 10 m/s. If the radius of the circle is 20 m, find the time period of the particle.

Answer: T = 2πr / v = (2 × 3.14 × 20) / 10 = 12.56 s

3 Question: A car is moving in a circular path with a speed of 30 m/s. If the radius of the circle is 100 m, find the centripetal force acting on the car.

Answer: $F = (m \times v^2) / r = (m \times 30^2) / 100 = 9m$ N

Unsolved Exercises
Conceptual Questions

1. What is the difference between a scalar and a vector quantity?

2. Can an object have a constant speed and still be accelerating?

3. What is the concept of centripetal force?

4. Can an object move in a circular path without any force acting on it?

5. What is the relationship between the radius of a circle and the circumference?

Short Answer Type Questions

1. Define the terms "velocity" and "acceleration" in the context of motion in a plane.

2. Explain the concept of centripetal acceleration.

3. Describe the motion of an object moving in a circular path with a constant speed.

4. What is the difference between a circular motion and a rotational motion?

5. Define the term "angular velocity".

Long Answer Type Questions

1. Derive the equation for the centripetal force acting on an object moving in a circular path.

2. Explain the concept of angular momentum and its relationship with torque.

3. Describe the motion of an object moving in a circular path with a variable speed.

4. What is the relationship between the radius of a circle and the period of revolution?

5. Derive the equation for the angular acceleration of an object moving in a circular path.

Numerical Problems

1. A car is moving in a circular path with a speed of 20 m/s. If the radius of the circle is 50 m, find the centripetal acceleration of the car.

2. An object is moving in a circular path with a constant speed of 10 m/s. If the radius of the circle is 20 m, find the period of revolution.

3. A particle is moving in a circular path with an angular velocity of 2 rad/s. If the radius of the circle is 15 m, find the linear velocity of the particle.

4. A car is moving in a circular path with a speed of 30 m/s. If the radius of the circle is 100 m, find the centripetal force acting on the car.

5. An object is moving in a circular path with a variable speed. If the radius of the circle is 25 m and the period of revolution is 4 s, find the average speed of the object.

Hints

- For questions 1-5, think about the fundamental concepts of motion in a plane and how they relate to each other.

- For questions 6-10, review the definitions of key terms and concepts in motion in a plane.

- For questions 11-15, use mathematical derivations and explanations to support your answers.

- For questions 16-20, use numerical values and equations to solve the problems.

More Questions

1. What is the relationship between the linear velocity and angular velocity of an object moving in a circular path?

2. Can an object have a zero linear velocity and still have a non-zero angular velocity?

3. Describe the motion of an object moving in a circular path with a constant acceleration.

4. What is the difference between a centripetal force and a centrifugal force?

5. Derive the equation for the centripetal acceleration of an object moving in a circular path.

6. An object is moving in a circular path with a speed of 15 m/s. If the radius of the circle is 30 m, find the centripetal force acting on the object.

7. A car is moving in a circular path with a speed of 25 m/s. If the radius of the circle is 60 m, find the period of revolution.

8. A particle is moving in a circular path with an angular velocity of 3 rad/s. If the radius of the circle is 20 m, find the linear velocity of the particle.

9. An object is moving in a circular path with a variable speed. If the radius of the circle is 40 m and the

period of revolution is 6 s, find the average speed of the object.

10. What is the relationship between the radius of a circle and the centripetal force acting on an object moving in a circular path?

Hints

- For questions 21-25, think about the relationships between linear and angular quantities in motion in a plane.

- For questions 26-30, use numerical values and equations to solve the problems.

Even More Questions

1. Can an object have a non-zero linear velocity and still have a zero angular velocity?

2. Describe the motion of an object moving in a circular path with a constant deceleration.

3. What is the difference between a circular motion and a spiral motion?

4. Derive the equation for the angular momentum of an object moving in a circular path.

5. An object is moving in a circular path with a speed of 20 m/s. If the radius of the circle is 50 m, find the angular momentum of the object.

CXXXX

Vectors to Velocity: A Guide to Motion and Mechanics By: Prashant Kumar Lal

www.ingramcontent.com/pod-product-compliance
Lightning Source LLC
Chambersburg PA
CBHW031613210526
45464CB00004B/1565